SpringerBriefs in History of Science and Technology

More information about this series at http://www.springer.com/series/10085

Allan Chapman

Mary Somerville and the World of Science

Springer

Allan Chapman
Wadham College and Faculty of History
Oxford
UK

ISSN 2211-4564   ISSN 2211-4572   (electronic)
ISBN 978-3-319-09398-7   ISBN 978-3-319-09399-4   (eBook)
DOI 10.1007/978-3-319-09399-4

Library of Congress Control Number: 2014947132

Springer Cham Heidelberg New York Dordrecht London

This book is a reprint of a title originally published by Canopus Publishing Ltd., 2004
ISBN 978-0953786848

Printed on acid-free paper

Springer is part of Springer Science+Business Media (www.springer.com)

Dedicated to the memory of the late Dame Kathleen Ollerenshaw, D.B.E. (1912-2014), mathematician, Fellow of Somerville College, and fellow-Mancunian.

Preface

It is always difficult to know which of those individuals who enjoy great prominence in their own time will go on to become part of the continuing cultural awareness of later generations. As scientists go, having one's name attached to a major discovery generally ensures survival, and those scientists who have been with us continually since the nineteenth century—such as Michael Faraday, Charles Darwin and Lord Kelvin—all have great discoveries to their credit. Yet who today, outside the domain of professional historical and scientific scholarship, knows about Sir John Herschel, William Whewell and William Buckland: men who made no world-changing discoveries, but who in their time commanded the highest international esteem as profound interpreters of science?

It is to this latter group that Mary Somerville (1780–1872) belongs. By the time of her death she basked in a reputation as an *interpreter* of science and of scientific culture (as distinct from a *populariser*) that embraced Europe and America, and her four major books were seen as masterly and original elucidations of the most complex scientific ideas of the age. Furthermore, she came to her conclusions not after a long and elaborate university education—for as a woman, all the normal channels of higher education were closed to her—but rather her achievement rests on the extraordinary tenacity and determination of her own self-education: from the secret girlhood reading of 'banned' mathematical books, which her well-meaning father believed would turn her mad, to her adult grappling with the highest branches of French mathematics.

Yet what was so conducive to Mary Somerville's career was the very openness of British science in the eighteenth and nineteenth centuries. Unlike the countries of mainland Europe, the creative heart of British science—especially in astronomy, the physical sciences and geology—lay not in formal national academies or state-funded universities and research institutes, but in the innovations and social relations of financially independent private scientists. These Grand Amateurs, or private 'lovers' of science, formed a remarkably open-ended community, in which if a person had originality and genius, the absence of a professorship or a university degree was of little consequence.

This world of private scientific gentlemen, with its sociable house-parties, dinners and soirees, could also be accessible to talented women, and a cultured and socially adroit person like Mary Somerville found relatively little difficulty in moving within it, especially as her work was fully supported by her second husband, Dr. William Somerville, F.R.S. This was, after all, the social world of Jane Austen, and later, Anthony Trollope. Imagine *Emma* and *Barchester Towers,* with the addition of science. It is a world which runs through this book.

Following her death in 1872, Mary's reputation as a female intellectual was such that her name was adopted by Oxford's new women's college, Somerville, in 1879. Yet though her name, portraits and papers were perpetuated and curated by the College, and she was revered as a female role-model, the detailed memory of her scientific achievements gradually slid from prominence. But given what has been said above about the nature of scientific fame, this was perhaps understandable; for there was no Somerville's Law or Somerville's Equation to keep her name in contemporary scientific textbooks. And as time passed, her brilliant interpretations of Laplace, physical science and geography slipped into obsolescence. She was clearly aware by her eighties, moreover, that the Newtonian and Laplacian celestial mechanics, which had stood at the very pinnacle of physics in 1830, were no longer at its forefront. The new science of astrophysics had come into being during the early 1860s, and the spectroscope offered a multitude of exciting new ways of expanding our knowledge of the Universe—such as the exploration of the chemical composition of stars and nebulae—beyond the exact calculation of orbital dynamics. In this respect, therefore, Mary Somerville's science had ceased to be in the ascendant, and one suspects that her subsequent historical neglect was related to this fact.

In 1873, the year after her death, her daughter Martha Somerville brought out the published version of her *Personal Recollections,* which was an edited and tidied version of the two manuscript autobiographical drafts that Mary had composed during the 1860s, and which are now preserved in the Bodleian Library, Oxford. In the published *Personal Recollections,* Mary emerges as a rather iconic figure—a role model for younger women to follow, and living proof of what intelligence and determination could achieve. She is always cool-headed, knows how to act whatever the circumstances, is energetic, and deals calmly with whatever life throws at her.

Yet while her daughter had carried out a certain amount of 'smoothing out' of her mother's life, one still finds, when looking at her manuscripts and private papers, that the woman who comes through them is essentially the same Mary Somerville as she whom one encounters in *Personal Recollections.* For she *was* a tough-minded, cool-headed and courteous individual with a pleasant disposition. She was also a great beauty whose looks survived well into middle age, as well as a woman who believed that 'femininity' counted for more than 'bluestockingism', and who loved dresses, theatre and balls just as much as she loved higher mathematics.

But it was the cruel and contemptuous treatment of the vulnerable and of the underdog that truly aroused her anger. Those men who were cruel to women, or else believed that women were uneducable, provoked a heated response, along with those who were cruel to animals, or to the wild birds, which she so loved. One of the most scathing passages in her published *Personal Recollections*, for instance, was reserved for the French anatomist Dr. François Magendie, for she 'detested the man for his wanton cruelties' practised upon the animals used in his physiological experiments (pp.192–193). On such occasions, one glimpses Mary Somerville's aroused passion, and can easily imagine what words and turns of phrase might have strayed from her lips. Yet those very traits which might have detracted from her posthumous role-model status in 1873 only make her more full-blooded and appealing to a modern reader.

Although Mary Somerville never lost her power to inspire intellectual young women—especially through the Oxford college which bore her name—it was not until the 1970s that serious research into her life and career began. The first notable scholarly treatise was the late Dr. Elizabeth Patterson's Oxford doctoral thesis, completed in 1980; and she also produced a 44-page monograph, *Mary Somerville 1780–1872* (Oxford, 1979), which, as part of Somerville College's centenary celebrations in 1979, surveyed the whole of Mary's career. Dr. Patterson then turned her doctoral thesis into a masterly book, bearing the title, *Mary Somerville and the Cultivation of Science, 1815–1840* (1983), which examines the formative and most intellectually significant years of Mary's scientific career before she and her husband and daughters 'retired' to Italy.

New insights into Mary's life were also provided in Dr. Mary Brück's article 'Mary Somerville, mathematician and astronomer of underused talents' *(Journal of the British Astronomical Association*, 106, 4 (1996), 201–206), in which it was argued that due to the circumstances of the age in which she lived, Mary Somerville was never able to achieve her obvious potential as an original research scientist. Then, in 2001 Sarah Parkin completed her 'Mary Somerville (1780–1872): Her Correspondence and Work in Chemistry', submitted as a one-year Chemistry Part II thesis for the Oxford University M.Chem. degree, and dealing with Mary Somerville's interests in photochemistry and spectroscopy.

In the same year, a new critical edition of *Personal Recollections* was published, with a prefixed title borrowed from one of Mary's obituaries of 1872: *Queen of Science. Personal Recollections of Mary Somerville* (Canongate Classics 102, Edinburgh, 2001). The editor of *Queen of Science* was Professor Dorothy McMillan, of Glasgow University's English Department, and with this book, Mary Somerville's *Personal Recollections* became easily available for the first time in well over a century. Its critical and biographical apparatus greatly amplifies Martha Somerville's edited text of her mother's autobiography, and its index further provides a merciful accessibility to a text which in its original 1873 printing had only an itemised title-page, in consequence of which it was often infuriatingly difficult to locate specific individuals and events. *Queen of Science* also restores those sections of autobiography in Mary Somerville's own hand, which Martha removed from her mother's original drafts, and which presented a less perfect image of the woman— such as her proving to be a very willing pupil when, as a child, being taught to swear by her army officer uncle (p. 9)—than a Victorian female icon-builder may have desired. These clearly marked sections, reincorporated from the original autobiographical drafts preserved in the Bodleian Library, Oxford, portray Mary as a more rounded and robust individual, though they in no way detract from her genuine likeableness. Perhaps what they show most of all are changing tastes in how different generations like to imagine their heroes and heroines: as carefully crafted icons conforming to type, or else as three-dimensional people, seen 'warts and all'. Even so, there are remarkably few warts on Mary Somerville. There was no scandal, no impropriety (apart from her beliefs, originally perceived as outrageous, about the capacities and potential of women), and a great deal of good nature and good company.

This book is in no way meant to be a definitive biography of this remarkable scientist. Instead, it is intended to make her life, achievements and scientific context available to a wider world, and hopefully, stimulate more scholars to undertake research into those original documents—including hundreds of letters—preserved in the Somerville Papers in the Bodleian Library, Oxford, and in the archives of the Royal Society.

My own interest in Mary Somerville really began in the early 1990s when I was researching my *Victorian Amateur Astronomer. Independent Astronomical Research in Britain, 1820–1920* (1998), in which her work is discussed. Then, in 1999 I was invited by Somerville College to speak on Mary Somerville at one of its Literary Lunches; and out of the interest generated from that lecture, delivered in June of that year, the present book came to be written. Reading through her letters, travel narratives, and those autobiographical drafts, which comprised her *Personal Recollections,* I became fascinated not only with the achievements and personality of Mary Somerville, but also with the scientific and broader social culture in which she lived and moved. It is my hope that the following pages will convey this fascination to the reader, help to place Mary Somerville into historical context, and to thereby interpret for a twenty-first-century audience the life of one of the greatest interpreters of science of the nineteenth century.

Acknowledgments

Many people have assisted me in various ways in the research and writing of this book. As a group, I thank my students in Oxford and elsewhere, upon whom I have tried out and developed my ideas on the Victorian scientific community over the years. Similarly, I express my gratitude to the staff of the Royal Society and Bodleian libraries, under whose care much of the Somerville collections and many of the letters are conserved.

As for individuals: I thank Liz Cook, the late Dr. Barbara Craig and the Principal and Fellows of Somerville College, Oxford, whose invitation to lecture on Mary Somerville to a College Literary Lunch in June 1999 formed the genesis of this book; Sarah Parkin, then of University College, Oxford, for the insight which I acquired when supervising her 2001 thesis on the chemical interests of Mary Somerville; the late Dr. Mary Brück, for her published scholarship; Peter Hingley and Mary Chibnall, of the Royal Astronomical Society library; and Bob Marriott, whose informed copy editing of the finished manuscript, and processing of the illustrations, was invaluable.

Above all, however, I am indebted to my wife Rachel, for not only typing my handwritten text and subjecting it to a thorough and critical read-through, but also for tolerating 'my other woman' during the process of research and composition.

Allan Chapman

Contents

Chapter 1
Mary Somerville and Her Influence

On 28 July 1897, Miss Florence Taylor, a young woman who was probably a school teacher by profession, delivered a lecture to the newly refounded Leeds Astronomical Society. It was entitled 'Mary Somerville, the great Woman Astronomer and Mathematician', and when published in the Society's *Transactions,* formed part of that genre of literature in which a person of a younger generation seeks inspiration and direction from the achievements of an illustrious forbear. By 1897, of course, higher education was gradually becoming more accessible to women, with the establishment of Royal Holloway, Girton, Somerville, and other colleges; though as Florence Taylor pointed out, such education had been virtually non-existent a century earlier when Mary Somerville had been young. Indeed, the increasing access to education which women had come to enjoy by 1897 was seen by Florence Taylor as yet another symptom of that driving force of nineteenth-century culture: 'Emancipation ... as witnessed by the Liberation of the Slave, (and the) Unfettering of the human mind from the thraldom of authority and prescription.' And as an Abolitionist, educational reformer, and hater of humbug in her own right, one suspects that Mary Somerville would in many ways have sympathised with Florence Taylor.[1]

Yet even in 1897, and in spite of growing educational opportunities, there were still pitifully few paid jobs for women in science and astronomy, especially in Britain and the rest of Europe. Apart from a small number of school and collegiate teaching posts, there was little beyond the short-term contract Assistantships at the Royal Observatory, Greenwich, paying £50 per annum, and Miss Annie Walker's permanent Second Assistantship at the Cambridge University Observatory with its £100 a year salary.[2] It is true that prospects were a little better in America. In 1865

---

[1] Florence Taylor, 'Mary Somerville, the Great Woman Astronomer and Mathematician', published lecture, 28 July 1897, in *Leeds Astronomical Society Journal and Transactions,* 5 (1897), 33–37.

[2] Cambridge University Observatory Syndicate *Report,* 1882. See Roger Hutchins, *A Mismatch of Ideals and Resources: British University Observatories c. 1820–1939* (Ashgate Press, Aldershot, 2002), Chapter II.

© The Author(s) 2015
A. Chapman, *Mary Somerville and the World of Science,*
SpringerBriefs in History of Science and Technology,
DOI 10.1007/978-3-319-09399-4_1

Maria Mitchell had become Professor of Astronomy and Director of the Observatory at Vassar College, and over the years she educated several generations of American astronomical women; although once they had their degrees, serious employment was still virtually non-existent beyond positions in a handful of colleges, or an Assistantship at Harvard Observatory, where Henrietta Leavitt, Annie Jump Cannon and others came to make significant discoveries when measuring and analysing photographic plates of star-fields.[3]

However, to think of nineteenth-century scientific women in terms of paid employment, or even as institutionally educated, is to miss the point as far as their effective careers is concerned, as this is a perspective forged by the historical circumstances not of the nineteenth-century but of the twentieth, and by all that has taken place both in terms of expectation and of actual achievement between 1901 and 2001. For in our own time, a 'career' in any sphere of human endeavour has become equated with the holding of a permanent salaried post in which prestige, creative opportunity, influence and pecuniary advancement are intimately bound together.

Yet in 1900 this was still by no means the only career path available to a scholar, whilst in 1850, or in 1800, a salaried post would have been the route adopted only by those who lacked such independent means as were necessary to be a private scholar or scientist. From the time of the Scientific Revolution in the seventeenth-century, and throughout most of the nineteenth-century, the status of private gentleman (or lady) gave one a greater cachet than being a person in receipt of a fixed salary with a board of trustees to obey.

This was a culture which might be called 'Grand Amateur': grand insofar as it aspired to do work of original, cutting-edge quality that would fundamentally advance science; and amateur, because, as Mary Somerville's scientific friend and encourager, Sir John Herschel, defined it, these people did science from *love*, and not from a pecuniary incentive.[4]

And of course, as one would expect, in the seventeenth and eighteenth centuries science was an overwhelmingly male culture. It was not for nothing, however, that it engendered a remarkably open and fluid world in which clergymen, physicians, schoolmasters, barristers and well-to-do private gentlemen could come together on equal terms to exchange ideas, compare instruments and observations, and publish their work. Cosmology, gravitational and celestial mechanics, planetary astronomy, and the complex motions of binary and triple stars—not to mention other sciences such as natural history and geology—were all advanced by this Grand Amateur culture. Charles Darwin, for instance, whose *On the Origin of Species* was cautiously admired by the 80-year-old Mary Somerville, never had a paid job in his

---

[3] Peggy Kidwell, 'Women in Astronomy', in *History of Astronomy. An Encyclopedia*, ed. John Lankford (Garland, New York, 1997), 564–567.

[4] For a wider insight into this 'Grand Amateur' tradition, see Allan Chapman, *The Victorian Amateur Astronomer. Independent Astronomical Research in Britain*, 1820–1920 (Praxis–Wiley, Chichester and New York, 1998).

entire life, and developed his work on natural selection on the strength of private inherited wealth.

This Grand Amateur scientific culture was, however, a peculiarly British, and perhaps American, phenomenon. Post-Civil War and Glorious Revolution Britain was concerned with keeping the monarch's wings well clipped when it came to wielding patronage, and Parliamentary vigilance—especially that of the Whigs— aimed at keeping taxation low and in the hands of the Commons. Even powerful aristocratic magnates, such as the Fifth Duke of Devonshire, were keen to limit the power of the King, while Britain's middle class—the largest and most politically self-aware of any European country—jealously guarded its wealth and independence.[5]

These social and political circumstances gave Britain a unique type of scientific association. Unlike that of France, Prussia, Austria, Russia and the German and Italian Ducal States, British science was not organised hierarchically, with a monarch presiding over a national academy. No powerful minister of state nominated approved *savants* for membership of a semi-courtly body, nor blessed them with funds and favour, as in Paris or St Petersburg. Quite contrarily, in fact, the Royal Society, founded in 1660, had been a symbol of monarchical poverty, for whilst King Charles II gave his *savants* a title, an impressive charter, and a golden ceremonial mace, its Fellows had been obliged to levy a weekly subscription upon themselves to pay for coal, candles and the Operator's wages. Yet this poverty became the key to their intellectual freedom, for they were left to pursue their investigations unhindered by the technical demands of war or statecraft.

Indeed, the Royal Society, paying its own way yet under the formal patronage of the Crown, modelled itself on the egalitarian Fellowship pattern of an Oxford or Cambridge College, electing its own members and deciding its own lines of investigation, rather than on what was the practice in an hierarchical academy. And this Fellowship form of association, run by its members, became the pattern for subsequent academic bodies in Britain, such as the Society of Antiquaries (1707), the botanical Linnean Society (1788), the Geological Society (1808), and the Royal Astronomical Society (1820). Likewise the Royal Academy for artists enjoyed the status of a Royal Charter (1768), but was dependent on the sale of the Academicians' pictures at its annual shows to remain in business and finance its teaching activities. All of them were prestigious, yet all were obliged to pay their own way, receiving no pecuniary assistance from the state. Self-funded and self-governed as they were, however, these societies set the scientific standards of the age, and formed the peer group within which the quality of research would be evaluated. The medals and citations of these societies, combined with the granting of honorary degrees from the English, Irish and Scottish universities, and the honours bestowed by the European academies, defined excellence in the international world of science that was in no way dependent upon holding a professorship or directing a research institute.

---

[5] The ongoing strategy of the Whigs to curtail monarchical power runs through Amanda Foreman's *Georgiana, Duchess of Devonshire* (Harper Collins, London, 1998).

The people who drew stipends for astronomical and mathematical work in Britain were few and far between. They included the two Savilian Professors, of Astronomy and Geometry, in Oxford, the Lucasian, Plumian and Lowndean Professors in Cambridge, the Andrews Professorship at Trinity College, Dublin, the often distinguished yet poorly-endowed Professors in the four Scottish universities, and the Astronomer Royal at Greenwich. There were also those men who 'professed' navigation or cartography at the Greenwich, Woolwich and Portsmouth officer training schools of the army, the navy and the East India Company at Haileybury. Yet even here, salaries in the way that we think of them today scarcely existed. Most of the incumbents of the Oxbridge scientific chairs, for instance, were generally beneficed clergymen, who often combined their academic posts with college livings and other pluralities, and also private means. Their social status derived from their being Master of Arts and members of clerical Convocation, just as if they professed Divinity, Ecclesiastical History or Hebrew, and not from being professional scientists per se. Ordained Astronomers Royal, such as the Revd Dr Nevil Maskelyne, were often similarly circumstanced, while all of these individuals, along with the armed forces professors, were by definition 'gentlemen' whose time, as well as most of their resources, was regarded as largely their own.[6]

This was the social and economic basis of the Grand Amateur world—a world in which private independence brought a higher kudos than did a paid job, and where the ticket of entry into the community of serious scientists depended more on talent, intellectual distinction, personal resources, and social compatibility within the 'club' of the learned than it did upon catching the eye of a powerful patron.

And perhaps most important of all—especially in the context of scientific women—acceptance within the English scientific world did not hinge upon the possession of formal academic qualifications. With the exception of a relative handful of Cambridge-trained mathematicians, most British astronomers and physical scientists in 1827—when Mary Somerville began to write *On the Mechanism of the Heavens*—were the products, to some degree, of substantial self-education, as far as their science was concerned. A young man might indeed hold an English, Scottish or Irish M.A. degree or a doctorate in Divinity, Civil Laws or Medicine, or be a Bencher in one of the Inns of Court; but that knowledge of physics or celestial mechanics that won him his F.R.S. was most likely to have been gained through assiduous private study some time after acquiring his predominantly arts qualifications. So many of the figures who were to become formative influences upon and encouragers of Mary Somerville came exactly from this tradition: Lord Henry Brougham, lawyer, scientific populariser, and later Lord Chancellor of England; Sir James South, the St Thomas's Hospital surgeon who owned the best-equipped observatory in England in 1830; Dr Thomas Young, practising physician, first decipherer of Egyptian hieroglyphs, and proponent of the wave theory of light; Sir Humphrey Davy, the self-educated Cornish apothecary's apprentice who became the most distinguished chemist of the early nineteenth-century; Dr Francis

---

[6] *Chapman*, The Victorian Amateur Astronomer [n. 4], 14–18.

Wollaston, physician and successful inventor, who made £30,000 from his process for the manufacture of malleable platinum, and who in addition discovered in 1802 those dark lines in the solar spectrum now known as 'Fraunhofer lines'. Wollaston, moreover, was a good friend of Dr Young, and also undertook researches which indicated that light is a wave phenomenon.

In continental Europe, however, the French Napoleonic and the re-founded post-Napoleonic Prussian education systems stressed science as a formal academic discipline. The École Polytechnique and the new universities of Berlin, Giessen, Dorpat, and many others, were laying the basis of the modern academic profession in science, with the Ph.D. degree as *de rigueur* for all who wished to be taken seriously and to make a career in science. Post-Congress of Vienna European governments after 1815, moreover, were not only lurching into reaction against the horrors of the French Revolutionary and Bonapartist eras, but their often alarmingly autocratic, newly-secured monarchies saw scientific patronage as one of the avenues through which intellectual as well as political authority could be expressed. A bright young man from a modest background could rise up through the great, usually Germanic, universities, obtain a Ph.D., a professorship, or an observatory directorship, and become an ornament of state. Friedrich George Wilhelm Struve, Friedrich Wilhelm Bessel, Johann Encke, to name but a few, all followed this career path between 1800 and 1830, while some were head-hunted to direct prestigious institutions abroad. For example, Friedrich G.W. Struve—the son of a North German Lutheran schoolmaster of peasant ancestry—became Czar Nicholas I's intellectual *coup* when he was appointed to the Directorship of the new Pulkowa Observatory, near St Petersburg, in 1839. Succeeding generations of the Struve family became minor Russian aristocrats,[7] and throughout five generations, seven members of the family were astronomers (The last of them, Otto Struve, was Director of Yerkes Observatory and later of Mount Wilson and Palomar Observatories, and died in 1964).

Observatories, laboratories and university science departments came to be occupied by a new scientific élite, with clear ladders of promotion, and precise understandings of who was (and who was not) a *real* scientist within that culture. Needless to say, it was a culture that differed fundamentally from that of the British Grand Amateurs—although the eclectically-trained English scientists were deeply respected in continental Europe, many of them receiving honorary degrees from its universities and medals from its academies.

---

[7] Alan H. Batten, *Resolute and Undertaking Characters: The Lives of Wilhelm and Otto Struve* (Reidel, Dordrecht, etc., 1988). See also Z.K. Sokolovskya, 'Friedrich G.W. Struve' and 'Otto W. Struve', *Dictionary of Scientific Biography,* ed. C.C. Gillespie, vols. 13–14 (Scribner's, New York, 1981), 108–121, for essay articles on individuals. A. Chapman, 'The Astronomical Revolution' (in nineteenth-century Germany), in *Möbius and his Band. Mathematics and Astronomy in Nineteenth-century Germany,* ed. John Fauvel, Raymond Flood and Robin Wilson (Oxford University Press, 1993), 34–77; also A. Chapman, 'The Professionalization of Astronomy in Nineteenth-Century Europe' (10,000 words), *Enciclopedia Italiana* , Vol. VII, *Storia della scienza* (Rome, 2003).

It would, however, have been much more difficult for a woman to win distinction
in continental Europe, especially in Germany, for the social complexion of German
science in particular, with its stress on research degrees and professorships, lacked
that essential component of easy fellowship which made it possible for Mary
Somerville to win the peer recognition as she did in the British Isles. On the other
hand, one should not forget that exceptional female intellectuals were not new.
Indeed, Mary's beloved Italy had in the eighteenth-century produced Maria Agnesi,
the daughter of a Bolognese Professor of Mathematics, whose brilliance was offi-
cially confirmed by Papal Decree, when she was authorised to lecture in the University when her father became ill.[8]

Mary Somerville, moreover, was not even the first woman to win fame and
recognition for her astronomical and scientific work within the English Grand
Amateur world. Nor was she alone in her own time, for Ada, Countess Lovelace,
Lord Byron's daughter, had impressed Charles Babbage and others with her
mathematical brilliance before dying at the age of 36 in 1852.[9] Yet both had been
preceded by Caroline Lucretia Herschel (1750–1848), a woman whom Mary
greatly admired, and whom she described as 'a lady so justly celebrated for
astronomical knowledge and discovery'.[10] Caroline Herschel was the sister of Sir
William Herschel, and at the age of 22 had been rescued from a life of domestic
drudgery in her parents' house in Hanover, by her brother who, by 1772, was
advancing his own musical career in Bath. He wanted Caroline to live as his
domestic companion, and when he realised that she was also well endowed with
Herschel musicality, he began to train her for a musical career of her own. At the
same time, William was using the quite substantial income of around £400 a year
generated from his musical activities to build telescopes and begin a programme of
serious astronomical research.[11] Finding that Caroline possessed scientific capa-
bilities as well as musical talent, he further trained her in astronomy, practical
optics, and observational techniques. And true to the Grand Amateur tradition, both
William and Caroline were entirely self-taught in science: from William Smith's
Compleat System of Opticks (1738) and other books, mainly in English—a language, indeed, that both of the German-speaking Herschels so quickly mastered that
in 1779 the Marchioness of Lothian complimented Caroline on 'pronouncing [her]

[8] See Ref. [1].

[9] *Doron Swade, The Cogwheel Brain. Charles Babbage and the Quest to build the First
Computer* (Little, Brown & Company, London, 2000), 155–171. Maboth Moseley, *Irascible
Genius. A Life of Charles Babbage, Inventor* (Hutchison, London, 1964), 155–66. *For a wider
history of the scientific interests of women, see Patricia Phillips, The Scientific Lady. A Social
History of Women's Scientific Interests, 1520–1918* (Weidenfeld and Nicolson, London, 1990),
207–209 (for Ada Lovelace and Mary Somerville).

[10] Mary Somerville, *On the Mechanism of the Heavens* (London, 1831), p.lxvi. Caroline Herschel
was also acknowledged in Mary Somerville, *On the Connexion of the Physical Sciences*, 3rd edn.
(London, 1836), Section XXXVI 'Fixed Stars', p.397. Patrick Moore, *Caroline Herschel,
Reflected Glory* (William Herschel Society, 1988).

[11] *Scientific Papers of Sir William Herschel*, I (Royal Society and Royal Astronomical Society,
1912). See Herschel's note for 1771, p.xxi, for income.

words like an Englishwoman'.[12] Caroline's surviving manuscripts, along with those of her brother, display a precision of grammar and syntax that in no way betrays the fact that English was indeed her second language. William seems to have also read French, and as a musician had a knowledge of Italian, though it is uncertain whether Caroline knew these languages (Fig. 1.1).

The scientific and broader international fame of the Herschels, however, was transformed in March 1781, when William discovered the planet Uranus—the first planet to be discovered since pre-Greek antiquity. Caroline was then working as her brother's astronomical assistant, and at the invitation of King George III, the Herschels gave up their musical careers in Bath to live in Slough to devote themselves full-time to astronomy. William was elected Fellow of the Royal Society, and received a £200 a year Royal Pension, though when he married in 1788 Caroline received £50 a year in her own right. And while this money was a Royal Pension in recognition of past achievements, and *not* a salary, it still made Caroline Herschel the first woman to receive a regular official income for her services to scientific research.[13]

While by the conventions of the day Caroline could not receive formal academic honours such as F.R.S., as conferred on her brother, she began to correspond on terms of equality with some of the leading physical scientists of the day. These included Dr Nevil Maskelyne, the Astronomer Royal; Sir Joseph Banks; President of the Royal Society; Charles Blagden and Alexander Aubert, both Fellows of the Royal Society and Grand Amateur scientists; and many more besides. Much of her work with these scientists concerned her own chosen line of astronomical research: cometography, which was different from that of her brother. As a practical astronomer in her own right, Caroline Herschel discovered eight comets, using a technique of 'sweeping' carefully selected zones of the sky with her own reflecting telescope. She was capable, using this telescope, of picking up comets when they were still a long way from the Sun and had not yet developed obvious tails. Her reporting of these discoveries to the Astronomer Royal, and to the Royal Society, from whom, of course, they were passed on to the international scientific community, enabled the approaches of these comets to the Sun to be exactly monitored over many months, and the geometrical shapes of their orbits to be measured precisely. This provided data from which it was possible to compute whether a comet would return to the Sun in a predictable closed orbit—such as Halley's comet, previously seen in 1758–1759 (and since, in 1910 and 1985–1986)—or else retreat into the depths of space. Comets, indeed, were key objects by which Newtonian gravitational theory could be physically tested, and Caroline Herschel was fully aware of their international scientific importance. In the 1790s she also undertook a major analysis of the *Historia Coelestis Britannica* (1725), the

---

[12] *Memoirs and Correspondence of Caroline Herschel*, ed. Mrs John Herschel (London, 1879), 40. The Herschel Papers, now preserved in Cambridge, show that from quite early in Caroline's life in England, she and her brother William privately corresponded in English.

[13] *Memoirs and Correspondence of Caroline Herschel* [n. 12], 50, 75.

Fig. 1.1 Miss Caroline Lucretia Herschel (1750–1848), sister of Sir William Herschel. She and Mary Somerville were made Honorary Members of the Royal Astronomical Society in 1835 (R.S. Ball, *Great Astronomers*)

catalogue of stars compiled by the Revd John Flamsteed, the first Astronomer Royal. (Her correspondent, Maskelyne, was the fifth holder of that office.) In spite of its age, Flamsteed's catalogue was then still the most comprehensive catalogue of the stars of the northern hemisphere, and Caroline's analysis found and corrected some 500 errors within it. Her corrections were published by the Royal Society in an elegant folio volume in 1798.[14]

---

[14] *Caroline Herschel, Catalogue of Stars from Mr Flamsteed's Observations contained in the second volume of the Historia Coelestis and not inserted in the British Catalogue* (Royal Society, London, 1798).

Fig. 1.2 Sir William
Frederick Herschel
(1738–1822), at about the age
when Mary Somerville met
him in 1816 (R.S. Ball, *Great
Astronomers*)

Mary Somerville made her first and what would turn out to be her lifelong
acquaintance with the Herschel family in 1816, when she and her husband Dr
William Somerville spent a day at the Herschels' Observatory House at Slough. The
Somervilles' Edinburgh friend Professor William Wallace had facilitated the
introduction, and the 78-year-old Sir William Herschel had shown them those
'celebrated telescopes' with which his 'numerous astronomical discoveries had
been made'. It was also on this visit that Mary met Sir William's 24-year-old son
John (later Sir John): 'quite a youth', who would later become 'my dear friend for
many years', and who would advise her, check her calculations, and strongly
encourage the publication of her *On the Mechanism of the Heavens*. Unfortunately,
on this 1816 visit Miss Caroline was 'abroad', or not present at Observatory
House.[15] In fact, Caroline had not lived there since her brother Sir William's
marriage in 1788, having taken lodgings in nearby Upton, and while brother and
sister had continued to work closely together, Caroline's main base had become her
own small come to graphic observatory near her lodgings (Fig. 1.2).

Numerous accounts of Caroline Herschel and her working relationship with her
older brother have been recorded, especially by scientists visiting England from the
Continent. When Barthelmy Faujas de Saint Fond arrived and was admitted by the

---

[15] *Mary Somerville, Personal Recollections from Early Life to Old Age of Mary Somerville. With
Selections from her Correspondence by her Daughter, Martha Somerville* (London, 1873), 105.

servant to the silent Observatory House one evening in August 1784, 'I observed, in a window at the farther end of the room, a young lady seated at a table, which was surrounded by several lights; she had a large book open before her, a pen in her hand, and directed her attention alternately to the hands of a pendulum-clock, and another dial placed beside her, the use of which I did not know; she then noted down her observations.' She explained to Saint Fond that her brother was working outside on a large telescope, and that through a system of strings (presumably working as a semaphore system) he could communicate to her the exact coordinates of the nebula or star cluster under observation. She then checked its celestial coordinates from Flamsteed's *Historia,* and its Right Ascension (east–west position) from the clock.[16]

In addition to her own cometary and other researches, Caroline spent more than 50 years 'reducing' her brother's 'deep sky' or cosmological observations, on past William's own death at the age of 84 in 1822, and after she had returned to Hanover after living 50 years in England. Her work would result in her becoming the first woman ever to receive a major international scientific award, when in 1828 the Astronomical Society of London (chartered as the Royal Astronomical Society in 1831) awarded her its Gold Medal. And 7 years later, in 1835, that same Society, the male pronoun in whose Royal Charter was to define the gender of its Ordinary Fellowship down to 1916, elected her an *Honorary* Fellow (Fig. 1.3).

In that same year, 1835, the Royal Astronomical Society also made Mary Somerville one of its Honorary Fellows, in recognition of her work in celestial mechanics and astronomical mathematics. Mary's election to the Royal Astronomical Society produced a letter of congratulation from Sir John Herschel, who in 1835 was continuing his late father's and aunt's cosmological surveys into the skies of the southern hemisphere. Following the conventions of the day, however, Herschel wrote from the Cape of Good Hope not to Mary direct, but to her husband, with the recognition that 'by a recent vote of the Astronomical Society I can now claim Mrs Somerville as a *colleague'*. This was a profound change of status for a scientific woman in 1835.[17]

It is not clear whether Mary Somerville and Caroline Herschel ever met, though as neither of their respective *Personal Recollections* or Diary *Memoirs* records such a meeting, either in England between 1816 and Caroline's departure for Germany in

---

[16] Barthelmy Faujas de Saint Fond, *A Journey Through England and Scotland to the Hebrides in 1784* (Paris, 1797), transl. Sir Archibald Geikie, (Glasgow, 1907), 63–65.

[17] Sir John Herschel to Dr William Somerville, 17 July 1830 (correct date, 1835), in Mary Somerville, *Personal Recollections* [n. 15], 217. This letter is clearly misdated, for Mary Somerville's election as Honorary Member of the Royal Astronomical Society did not occur until 1835: see Mary Somerville to Augustus de Morgan, 20 February 1835, expressing her pleasure at being so elected: R.A.S. Letters, 1835. Moreover, Herschel's address on his letter of 17 July 1835 —'Feldhausen', Cape Town—was his residence between 1834 and 1838. In July 1830 he was living in England, not 'Feldhausen', Cape Town.

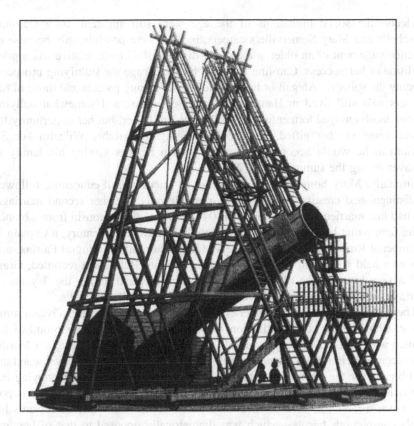

Fig. 1.3 Sir William Herschel's great telescope at Slough, built in 1788, and still the largest telescope in the world when Mary and William Somerville saw it in 1816. Its mirror was 48 inches in diameter, and its tube was 40 feet long (Author's collection)

1822, or on the Continent between 1822 and 1848, it seems improbable. However, Mary's admiration for Caroline Herschel was very clear. On 16 April 1835, Mary wrote to her older colleague saying that her own recent election to the Royal Astronomical Society came with especial 'distinction by associating my name with yours, to which I have looked up with so much admiration'. She also presented Miss Herschel with a copy of her recent *On the Connexion of the Physical Sciences* (1834), 'which is offered with great deference, having been written for a very different class of readers'.[18]

While the more fluid world of British Grand Amateur science made it possible for two women to rise to its very pinnacle of recognition—not to paid Professorships, but to honoured *colleague* status within a Society of independent original

---

[18] Mary Somerville to Caroline Herschel, 16 April 1835, reproduced in *Memoirs and Correspondence of Caroline Herschel* [n. 12], 274.

intellects—the social limitations of the age were still apparent. Both Caroline Herschel's and Mary Somerville's careers, indeed, were possible only because of the encouragement of an older and already distinguished male relative. As a prior condition of her success, Caroline Herschel had to escape the stultifying prospects of being the spinster 'Abigail or housemaid' to her ageing parents and those of her brothers who still lived in Hanover.[19] Caroline's eldest and somewhat bullying brother Jacob enjoyed tormenting and talking down to her, but her opportunity for release came via her gifted, humane and far-sighted brother Wilhelm—or Sir William as he would become in England—when he was visiting his family in Hanover during the summer of 1772.

Similarly, Mary Somerville's great surge of scientific self-education, followed by distinguished creative work, became possible only after her second marriage. She had first married Samuel Greig in 1804. He was a distant cousin from a branch of the family that had settled in Russia in the mid-eighteenth-century, a Captain in the Imperial Russian Navy who had served under her father, Admiral Fairfax, and who now held a Russian diplomatic post in London. But as she recorded, Greig 'had a very low opinion of the capacity of my sex', and during the 3 years of marriage prior to his death, he never encouraged her intellectual gifts.[20]

Then in 1812 she married another cousin, Dr William Somerville, a senior army doctor, who in clear contradistinction to Samuel Greig was openly proud of his talented wife, and did all that he could to help her. Indeed, as their daughter Martha later recorded, 'He was far happier in helping my mother in various ways, searching the libraries for the books she required, indefatigably copying and recopying her manuscripts, to save her time. No trouble seemed too great which he bestowed upon her; it was a labour of love'.[21] Indeed, William Somerville's attitude towards his wife's conspicuous talents—which was diametrically opposed to that of her first husband Samuel Greig—was of the greatest importance to the development of Mary's career, for 'The warmth with which Somerville entered into my success deeply affected me; for not one in ten thousand would have rejoiced at it as he did'.[22] Dr Somerville was a widely travelled, cosmopolitan and urbane Scottish physician who had seen service in South Africa and other distant places. But following the scaling down of the Army Medical Department after Waterloo in 1815, he was faced with the prospect of redundancy, and it is probable that his and Mary's continental tour of 1817 was occasioned partly by the relative cheapness of life in France and Italy in contrast with that in London, and partly by the prospect of practising medicine amongst the expatriate British (and particularly English) who went to live abroad. Then in 1818 William Somerville, who was a member of the Army Medical Board, was appointed to the post of Physician to the Royal Hospital,

[19] *Memoirs and Correspondence of Caroline Hersche* 1 [n. 1], 75.
[20] Mary Somerville, *Personal Recollections* [n. 15], 75.
[21] Mary Somerville, *Personal Recollections* [n. 15], 84.
[22] Mary Somerville, Personal Recollections [n. 15], 176.

Chelsea, and from this prestigious position, with all its natural connections into the most distinguished circles of metropolitan science and medicine, Mary's own career was ideally placed to take off.

Reference

1. Agnesi, M. G. (1718–1799). *Encyclopaedia Britannica*, (pp. 1929–1930). I, 350 XIV edn. London and New York.

Chapter 2
Early Life, Career and Friends: The Social World of Georgian Science

Mary Somerville had many blessings. In addition to her obvious intellectual gifts, she had 92 years of active life, an elegance of form which she retained into old age, and a beauty of disposition and a sympathy that gave her the gift of friendship. She was also practical and tough-minded, with no time for woolliness or pretence: a hard-headed individual in a romantic age.

Indeed, she was succinctly described by her fellow scientific authoress Maria Edgeworth, who in January 1822 wrote: 'Mrs. Somerville is the lady who La Place says, is the only woman in England who understands his works. She draws beautifully; and while her head is among the stars, her feet are firm upon the Earth. Mrs. Somerville [is] slightly made, fair hair, pink colour, small grey round, intelligent smiling eyes; very pleasing countenance; remarkably soft voice, strong but well-bred Scotch accent...'.[1]

As her *Dictionary of National Biography* recorder, Ellen Mary Clerke, expressed it more than 60 years later, her 'soubriquet of the Rose of Jedburgh formed a piquant contrast to her masculine breadth of intellect'. An interesting observation indeed, made by one late Victorian intellectual woman on a predecessor of an earlier generation. Yet one of the sources of Mary Somerville's success, like that of Caroline Herschel, was her instinctive sense of how to succeed in a man's world: a socially flexible and reputationalist world, yet one in which men still undoubtedly made the rules as far as the recognition of intellectual excellence was concerned.

Mary Somerville was born in The Manse, Jedburgh, on the Scottish borders, on 26 December 1780. Her father was Sir William George Fairfax, an honoured, distinguished, yet rather impoverished admiral in the Royal Navy. The Fairfaxes

---

[1] Maria Edgeworth to Miss Ruxton, 17 January 1822. Copy of letter in Somerville Papers, Bodleian MS. Dep.c. 370. MSB. 3–34, Collection 'E'. This same letter is also printed in Mary Somerville, *Personal Recollections* (London, 1873), 156, though it is headed 'Maria Edgeworth to Miss...', without any reference to Miss Ruxton. The reason for the omission of Miss Ruxton's name in the printed text is uncertain, as it is clearly spelt on the above-mentioned Bodleian Library document. As this document, Dep. c. 370, seems to be an early nineteenth-century copy of the original letter, however, the omission may possibly be due to an error of transcription.

© The Author(s) 2015
A. Chapman, *Mary Somerville and the World of Science*,
SpringerBriefs in History of Science and Technology,
DOI 10.1007/978-3-319-09399-4_2

were of Yorkshire origin, and claimed descent from Lord Thomas Fairfax, the Parliamentary General in the Civil War. Yet Sir William's victories had not been especially profitable, and whatever prize money he may have won had long since been depleted, and the family seems to have lived on his modest pension. During his service in the American station during the Wars of Independence, however, Sir William had struck up a correspondence friendship with George Washington, with whom he was related, and he always regretted that the exigencies of war had never allowed him to accept Washington's cordial invitation to visit him socially.[2]

Mary's mother was formerly Miss Margaret Charters, the daughter of a Scottish solicitor. Mary's background, apart from the naval and legal professions of her parents, consisted of Scottish clergy, academics, and minor gentry. And though Jedburgh-born, she always regarded Burntisland, a small fishing village on the Fifeshire coast, to be her real home.

When she was born in December 1780, in fact, Mary's mother had only just returned from waving her husband off to sea on a series of adventures which would not permit him to return until Mary was a girl of eight or nine—by which time she had been allowed to run wild and had received no regular education, though her freedom to explore the area around Burntisland where she spent her childhood gave her a healthy constitution and stimulated her intellectual curiosity in the natural world.[3] At this stage in her life she also showed signs of that robustness of mind and spirit that she would carry (in a more circumspectly expressed form) for the rest of her days. 'About this time I was with my mother on a visit to her father in Edinburgh when my uncle Thomas Charters an officer in the Indian army then on leave, amused himself by teaching me to swear. One day walking with my maid in the High Street a lady asked my name and I answered, 'What's your business you damned B****'. The lady said, 'You're a bonny bairn but weel awat ye hae an ill tongue'. It says something for the octogenarian Mary's sense of humour that she included this incident in her autobiographical draft, although her daughter Martha removed it from her *Personal Recollections* published in 1873.[4]

Following her father's return home from sea, however, Mary began to receive the usual education of a young lady of her age and social station at a private school in Musselborough, though from an early age she 'resented the injustice of the world in denying all those privileges of education to my sex which were so lavishly bestowed upon men'. She subsequently recorded, however, that her interest in

---

[2] Mary Somerville, *Personal Recollections* [n.1], 227.

[3] Elizabeth C. Patterson, *Mary Somerville and the Cultivation of Science, 1815–1840* (Martinus Nijhoff, Kluwer, Boston, The Hague, Lancaster, 1983), 1. Elizabeth C. Patterson, *Mary Somerville* (Oxford, 1979), 7–9. Mary Somerville, *Personal Recollections* [n.1], 20.

[4] This passage about childhood swearing appears under Mary Somerville's own hand in the First Autobiographical draft: Bodleian Library, Somerville Papers Dep. c. 355 MSAU-3, p. 5, but a later editor (probably Martha Somerville) has cut the entire passage out of the page with scissors, leaving a 5 × 4-inch hole in the sheet. The offending passage was restored by Dorothy McMillan (ed.), *Queen of Science. The Personal Recollections of Mary Somerville* (Canongate Classics 102, Edinburgh, 2001), 9.

mathematics was first aroused when, as a young girl, she was looking at a monthly fashion magazine with an older friend, Miss Ogilvie of Burntisland. The magazine contained a puzzle which Miss Ogilvie described as 'a Kind of Arithmetic: they call it Algebra; but I can tell you nothing about it'.[5] On one level, this story begs a whole host of questions: was there a sufficiently large number of young Scottish women who, in spite of Miss Ogilvie's personal ignorance, enjoyed tackling algebraic puzzles as part of their fashion reading? If not, how did such puzzles find their way into ladies' magazines in the first place? On the other hand, such puzzle articles in women's magazines must not have been uncommon, for it was in response to one of them—dealing, in this instance, with the mathematical vibrations of a musical string under tension—set in *The Ladies' Diary* for 1780, that resulted in Sir William Herschel's first publication.[6] But in any case, the puzzle article discovered in the fashion magazine led Mary Somerville to start searching for information about algebra. She obtained a copy of Robertson's *Navigation,* but lacked the background to understand it, especially as her arithmetic at this stage was so poor that she could never produce the same answer twice when she added up a simple column of figures. Yet the puzzle seems to have launched her into what would become a lifelong odyssey of largely self-taught scientific investigation.

Mary's next strategy was to persuade her Edinburgh student brother Henry to purchase books for her. She obtained a copy of Euclid's *Elements of Geometry,* and Bonnycastle's *Algebra,* in addition to which she set about learning Greek so that she could read Xenophon and Herodotus in their original language. In Edinburgh she had lessons in the more acceptable ladylike skills of art and drawing. Her tutor was Alexander Nasmyth, the founder of the Scottish landscape school, and he later remarked that 'the cleverest young lady [he] ever taught was Miss Mary Fairfax'. Nasmyth's teaching inspired Mary with a fascination with perspective and the mathematical structures within art, and, as Maria Edgeworth mentioned, she herself became an accomplished lifelong painter and draughtswoman. Alexander Nasmyth had another important connection with Mary Somerville, for in addition to being a successful artist he moved on equal terms amongst the scientists and medical men of the Scottish Enlightenment.[7] Sir David Brewster—the eminent optical physicist, and the discoverer of polarisation—was one of his circle, while Alexander's son James Nasmyth was to become one of the leading lights of the English Grand Amateur astronomical community from the 1840s to 1880s.[8]

Yet Mary's intellectual precocity alarmed her elderly father, who feared for his beloved daughter's health and sanity if she persisted in these demanding pursuits. Indeed, he said to his wife, Mary's mother: 'Peg, we must put a stop to this, or we shall have Mary in a strait jacket one of these days. There was X, who went raving

---

[5] Mary Somerville, *Personal Recollections* [n.1], 47.

[6] See [1].

[7] See [2, 3].

[8] *The Home Life of Sir David Brewster, by his daughter Mrs. Gordon* (Edinburgh, 1869). Also *James Nasmyth* [n.7], 63–95.

mad about the longitude'.[9] Mary then started studying her mathematical books in secret, late at night, after being put to bed. But the servant noticed that her night candles were mysteriously wasting away, as a result of which the candle was taken away once she was put to bed, leaving her with no source of light to read by. Not to be deprived of her beloved mathematical and linguistic studies, however, Mary decided to use the time for memory training, so she began to recall in detail all that she had been allowed to read during the day. Early on in life, Mary had become 'intensely ambitious to excel in something, for I felt in my own breast that women were capable of taking a higher place in creation than that assigned to them [which] in my early days was very low'.[10]

Being a woman of her age, and strongly respectful to her parents, who were now struggling to make ends meet on Sir William's miserly Royal Navy pension of £75 per annum, and also because she was strikingly attractive—and, as her daughter Martha later recorded, in no way a 'blue stocking'—Mary took the natural step of getting married at the age of 24. Her husband was the above-mentioned Captain Samuel Greig, and the marriage gave her a surviving son, Woronzow, and left her a widow after 3 years.[11]

In widowhood Mary returned to mathematics—in particular, Ferguson's *Astronomy* and Newton's *Principia*. She soon became aware, however, that while Newton had created a gravitational mechanics and new methods of computing the orbits of 'three bodies'—the Sun, Moon and Earth—and their mutual effects upon each other, this had been done 120 years earlier. A growing reverence for Newton had, unfortunately, placed something of a dead hand upon English mathematics in the interim, and superior calculating techniques such as the calculus (devised by Newton's German arch-rival Gottfried Leibniz) were virtually ignored in England —and perhaps nowhere more so than in Newton's own university of Cambridge.

It was in continental Europe, and especially in Paris, that most of the mathematical innovations of the previous 40 years had taken place. None of Newton's work was challenged or in any way undermined in this process, for French mathematicians such as Lagrange, La Croix and Laplace all agreed that Newton's genius still provided the bedrock on which all subsequent mathematical physics had been built. Yet the French mathematicians recognised that further developments had to take place, and new insights had to be obtained. One could not simply draw a reverential line under Newton's *Principia* and say 'enough'.

Mary Somerville belonged to the first generation of British mathematical scientists to recognise the need to shake off this torpor. Her subsequent friends Sir John Herschel, Charles Babbage, Augustus de Morgan, William Whewell and George Peacock—at that time still undergraduates or else young dons at Cambridge —set about the revival of Cambridge mathematics: brilliant junior members of the

---

[9] Mary Somerville, *Personal Recollections* [n.1], 54.

[10] Mary Somerville, *Personal Recollections* [n.1], 60.

[11] E. Patterson, *Mary Somerville* [n.3], 11–12. E. Patterson, *Mary Somerville and the Cultivation of Science* [n.3], 5.

University trying to arouse and inspire their own superiors. In 1812, indeed, they founded the Analytical Society, with the avowed intention, in spite of the war that was raging between the two countries, of opening up Cambridge to the creative mathematical work going on in France. Between them, these men were laying the foundation for Cambridge's excellence in the physical sciences during the Victorian age—a laurel indeed previously held by Oxford, when between 1695 and 1795 David Gregory, Edmond Halley, James Bradley, Nathaniel Bliss, Thomas Hornsby and others had pioneered both Newtonian physics and observational astronomy in Oxford University.[12]

Edinburgh University was also sensitive to the work of contemporary French mathematical scientists, and when Mary Somerville won an inscribed prize medal for solving a mathematical puzzle set by Dr. William Wallace (who in 1819 would become Edinburgh University's Professor of Mathematics), she made her first formal acquaintance with the world of academic science. She so impressed Professor Wallace, moreover, that he offered her further instruction, and a detailed reading list —most of the volumes upon which were in French. She purchased 'an excellent little library', encapsulating the most advanced mathematical thought of the day. And by this stage, it had become impossible to conceal her intellectual passions, though 'I was considered eccentric and foolish, and my conduct was highly disapproved of by many, especially by some members of my own family'.[13] Even so, it seems that she had her second husband Dr. William Somerville's fullest support.

At Wallace's suggestion, Mary Somerville embarked on the study of Pierre-Simon Laplace's *Mécanique Céleste,* not all of the five volumes of which were yet published. She was reading Laplace with William Wallace's mathematical brother, John, and rapidly found that she understood it as well as he did, which added to her confidence and spurred her on.

In 1815, *Mécanique Céleste* (1799–1825) was the most advanced statement on one of the most complex branches of mathematics ever to have been written. Building upon Newton's *Principia* (1687), and that mathematical and conceptual language which made it possible to express the exact and ever-changing gravitational interactions between the masses of the Sun and the planets as they moved in space, Laplace addressed the most complicated problems in contemporary cosmology. Why, for instance, did the orbits of Jupiter and Saturn seem to be changing over the centuries? Would Jupiter one day crash into the Sun, or would Saturn break free and spin out of the Solar System in millions of years to come? How far did the drag effect of the tides on the Earth affect our planet's, and the Moon's, long-term orbital relations? Would the Moon eventually break free from the Earth's gravitational pull? Indeed, how dynamically stable was the Solar System? Did the distant binary stars,

---

[12] For the mathematical innovations, see Ivor Gratton-Guinness, 'The young mathematician', in *John Herschel, 1792–1871: A Bicentennial Commemoration,* ed. D.G. King-Hele, F.R.S. (Royal Society, London, 1992), 17–28. For Oxford's astronomical excellence, see A. Chapman, 'Oxford's Newtonian School', in *Oxford Figures. 800 Years of Mathematics,* ed. John Fauvel, Raymond Flood and Robin Wilson (Oxford University Press, 2000 [ed. 2]?), 137–149.

[13] Mary Somerville, *Personal Recollections* [n.1], 80.

Fig. 2.1 Baron Pierre Simon Laplace (1749–1827), the great French cosmologist and mathematician whose researches inspired Mary Somerville to write *Mechanism of the Heavens* (R.S. Ball, *Great Astronomers*.)

which William Herschel discovered between 1782 and 1802, and more of which other astronomers later discovered, operate under the same gravitational laws as those which governed the Sun and planets? Could the co-planarity, and the same orbital direction, of all the planets around the Sun be used to prove that the planets had once been part of the Sun, from which they had been ejected only to condense as spinning spheres; and had the Sun once rotated faster than it does today? Laplace, indeed, had taken up some of the issues pertaining to the origin of the Solar System, and as early as 1796 had proposed his famous nebular hypothesis—or condensation theory—for the formation of the planets around the Sun, when he published his highly influential *Exposition du Système du Monde* (Fig. 2.1).

By the 1820s Mary was winning attention due to her obvious genius for higher mathematics. In addition to William and John Wallace, she was also encouraged by Professor John Playfair of Edinburgh, Dr. Thomas Young and Sir John Herschel in London, and the influential Henry, later Lord Henry, Brougham.

Brougham was an Edinburgh lawyer whose talent was rivalled only by his eccentricity and ambition. In the highest traditions of the Scottish Enlightenment, he was interested in all branches of humane learning, and was one of the leading reform politicians of the day. His reformist principles had made him too hot for conservative Edinburgh, however, and by 1808 he had moved south, been admitted to the London Bar, and entered Parliament 2 years later. Closely associated with the radical Whigs in Parliament, and sensitive to the more positive currents which had emerged out of revolutionary Europe, in 1830 he received a peerage and the Lord Chancellorship, whereupon he threw all of his exotic brilliance behind the passing of the great Reform Act of 1832. Brougham was, in addition, a firm believer in science as an engine of fundamental reform, and had already played a leading role in the establishment of the Mechanics' Institutions movement, which aimed to teach the principles of science to the working classes. He had also been greatly impressed by Mary Somerville, with whom he had probably become acquainted via Mary's friendship with Brougham's sister. And while not a particularly political person herself, one can see how Mary's own progressive views were in keeping with those of some of the radical Whigs, on issues such as Parliamentary reform, the education of the working classes (and later, the education of women), and the final abolition of slavery, which came in 1834. In these respects, she also shared common ground with the Edinburgh Professor's daughter Margaret Brodie Stewart, who in March 1829 married John Herschel.[14]

In short, Mary Somerville was a product of that remarkable and brilliant era called the Scottish Enlightenment, when Edinburgh became perhaps the most intellectually illustrious city in Europe. Though many of its leading figures held chairs in the university, the Scottish Enlightenment was an extraordinarily open movement in which talent, rather than birth or academic office, opened the doors. Even many of the men who held professorships had risen from poor backgrounds to international reputations. And its drawing-room, or club, culture gave full expression to those who were not post-holding academics, embracing people as diverse as the poet Robert Burns, Dr. Robert Knox the anatomist (who ran a distinguished private medical school, quite separate from the University, and bought cadavers from the notorious body-snatchers Burke and Hare), and Thomas Henderson, the one-time lawyer's scrivener who became Director of Edinburgh's Royal Observatory. Although this was a predominantly male culture, its openness made it possible for a woman of extraordinary talent to find both encouragement and creative expression, even when that woman had no starry illusions about the barriers which still existed in the Edinburgh world, and was all too aware of the conservatism and petty snobbery that still bedevilled the 'Athens of the North'.

Mary Somerville's intellectual connections within metropolitan science can be traced from the time of her move from Edinburgh to London after 1816, and her

---

[14] Lady Margaret Brodie Stewart Herschel makes many references in her correspondence to the abuses to which newly-freed slaves were subjected at the Cape of Good Hope in 1834: *Lady Herschel's Letters from the Cape, 1834–1838*, ed. Brian Warner (Friends of the South African Library, Cape Town, 1991), 43, 50, etc.

preserved correspondence and related documents read like a *Who's Who* of late Georgian science. On the other hand, a review of these names clearly reveals where her predominant scientific interests lay, for they were on the whole the names of physical rather than natural history scientists. In spite of William Somerville's profession, there are not many medical men, botanists or zoologists, and those who are to be found—such as Dr. Thomas Young and Dr. William Wollaston—are usually there because of their separate interests in physics or astronomy. The only natural history scientists whose names are prominent—such as Sir Charles Lyell, Sir Roderick Murchison, Adam Sedgwick, and the Revd Dr. William Buckland— are included because they are geologists. Many of these men, moreover, were married to intellectually gifted wives, with whom Mary formed lifelong friendships.

Mary was fascinated by the new and continuing discoveries which suggested that the Earth was immensely old rather than having been created only around 4004 BC, though in many ways what really interested her was the idea of the Earth as a planet whirling through space and condensing from its primordial Laplacian neb- ulosity, rather than the successive *genera* of living creatures which later came to populate it. Her books can be examined in vain for anything beyond passing references to organic fossils, in spite of their central significance to contemporary geologists; while her last book, *Microscopic and Molecular Science* (1869), is more concerned with the structural forces that might underlie living creatures than with the matters of habitat or ecology which would have interested a naturalist. (More will be said of these interests in Chap. 4.)

Even so, Mary moved easily amongst the leading fossil geologists of the day, and in February 1829 she and Dr Somerville spent an enjoyable week at Christ Church, Oxford, along with Sir Roderick and Lady Murchison, as guests of Mary and William Buckland. Buckland and Murchison were two of the leading geologists of the day, yet while Buckland held the Regius Chair in geology at Oxford in conjunction with a Christ Church Canonry, and Murchison was to become Director General of the Geological Survey, both men had risen to scientific prominence through the self-taught Grand Amateur tradition, the former being a clergyman, the latter a retired army officer. Buckland, moreover, was one of the foremost scientific celebrities of the time—a spell-binding speaker, a much sought-after dinner-party guest, and sometimes an outrageous eccentric, although none of these traits pre- vented his translation to the Deanery of Westminster in 1845. His Christ Church Canon's residence—a large free-standing house just off the Cathedral cloister— contained one of the world's finest private geological collections and a menagerie of exotic living creatures, so that it is impossible to imagine that natural history did not figure prominently in that week's conversation.[15] Charlotte Murchison and Mary Buckland, moreover, were intellectual women in their own right, and the three couples cemented friendships that would endure for the rest of their lives.

---

[15] Mary Somerville, *Personal Recollections* [n.1], p.30, for Oxford visit, though she does not give a date. February 1829 is given in E. Patterson, *Mary Somerville and the Cultivation of Science* [n.3], 53.

It is clear, however, that Mary was also much admired by Buckland's Cambridge counterpart, the Revd Dr. Adam Sedgwick, who was Woodwardian Professor of Geology in the University, as well as a Prebend of Norwich Cathedral, for the Somervilles stayed with him in Trinity College in April 1832. Sedgwick, who was a bachelor and lived for most of the year in College, except when off geologising or else fulfilling his residence at Norwich Cathedral, was also something of a larger-than-life figure and a social lion. Following the conventions of the day, he wrote to Dr. Somerville rather than to Mary prior to their visit to Cambridge, pointing out how the stern severities of a bachelor college were being softened somewhat in preparation for the couple's visit: A four-posted bed (a thing utterly out of our regular monastic system) will rear its head for you and Madame' in the set of rooms in which they would be staying. And almost as a way of reassuring the Somervilles that Trinity College's bachelor dons were capable of making a visiting lady comfortable, Sedgwick informed Dr. Somerville that a similar arrangement had worked excellently when Sir Roderick and Lady Charlotte Murchison had visited Trinity College. Sedgwick further asked whether on Mary's arrival in Cambridge she was likely to be exhausted from travel; although, as her husband pointed out, travel 'rather recruits Mrs. Somerville', and she was willing to go along with whatever had been arranged.[16]

In spite of Sedgwick's corresponding with William Somerville, it is clear that Mary was the real celebrity, for on 5 April Sedgwick unveiled the busy social schedule proposed for the visit, with receptions for people to meet Mary, and a succession of dinners hosted by the mathematicians William Whewell, George Peacock, George Airy, and the anatomist Dr. Peacock, on successive evenings. At the end of the week in Cambridge, Sedgwick and Whewell accompanied the Somervilles for several more days to Audley End. After their return to London, Mary wrote to Sedgwick to thank him for the splendid visit, emphasising her awareness of the recognition awarded for her scientific work by 'such men as adorn your University'.[17]

Indeed, the extremely social character of the scientific world of Mary Somerville's day is clearly evident from her letters and reminiscences. This was an age when transport was relatively slow—she was in her mid-sixties by the time that railways had begun to shrink distances across England—and the age of the academic conference as a shop window for state-of-the-art research was still some decades away. And while it is true that the annual week-long jamborees of the British Association for the Advancement of Science—with their dinners, balls, and sociable field trips—moved

---

[16] Dr. William Somerville to the Revd Prof. Adam Sedgwick, April 1832, in *The Life and Letters of the Reverend Adam Sedgwick LL.D. D.C.L., F.R.S.*, vol. 1 (Cambridge University Press, 1890), 388.

[17] Mary Somerville to Sedgwick, Chelsea, 25 April 1832: *Life and Letters of… Sedgwick* [n.16], 389. For Kater's work on the physics of vibrating pendulums and their experimental applications, see Henry Kater, 'An Account of Experiments for Determining the Length of the Pendulum vibrating Seconds in the Latitude of London', *Philosophical Transactions of the Royal Society*, 108 (1818), 33–102.

around the British Isles with great success after 1831, one of the most important instruments of scientific association remained—as it had been since the founding of the Royal Society in 1660—a meeting of friends—and rivals! For male scientists, these meetings could take place at the Royal Society or other metropolitan learned societies, although the soirees of the Royal Institution and the Surrey Institution, both of which admitted ladies to their gatherings, widened the scope for more formal contacts within London. Even so, as one quickly learns from Mary Somerville's private writings, social gatherings at private houses were an enormously important agent of scientific exchange, especially if ideas were aired across congenial dinner-tables or drawing-rooms.

One such incident—undated but probably occurring in the early 1820s—is recorded in the *Personal Recollections*. Mary and William Somerville had spent an evening with Captain Henry Kater and his wife Mary Frances. (Kater was the ex-army physicist who in 1818 had conducted pioneering researches into the behaviour of free-swinging pendulums, and invented in the 'Kater free pendulum' an instrument which would revolutionise geophysics by making it possible to measure and map slight variations in the Earth's gravitational field.)[18] After a musical interlude in which Captain and Mrs. Kater sang 'very prettily', and after much scientific discussion and the trial of several experiments, the party adjourned to the garden with at least one astronomical telescope. It was a clear night, and they proceeded to test the telescopes to determine their resolving power in the separation of double stars. This went on 'till about two in the morning', when the group noticed a light still burning at the window of Dr. Thomas Young's house nearby. Dr. Somerville rang the doorbell, and when Young, wearing his dressing gown, answered it, he invited the party in, and proceeded to show them a piece of Egyptian papyrus which seemed to have a Ptolemaic horoscope drawn upon it, and in which he was in the process of translating.

In addition to his work as the physicist who established the key experimental proofs for the wave theory of light, and serving as Professor of Natural Philosophy at the Royal Institution, Thomas Young made significant contributions to the decipherment of the ancient Egyptian hieroglyphic language. Indeed, Mary Somerville recorded that Jean-François Champollion, who finally deciphered the language in 1828, was 'ungenerous' insofar as he failed to acknowledge Young's contributions.[19]

---

[18] No date for the evening with the Katers is given in Mary Somerville, *Personal Recollections* [n.1], 130; nor is there any reference to the occasion in George Peacock's *Life of Thomas Young M.D., F.R.S., & C.* (London, 1855). It probably took place some time just before or after 1820, however, when Young was doing his work on the ancient Egyptian language and script. At that time (c.1801–1826) he was living at 48 Welbeck Street, London: *Life*, 468 and 253. Unfortunately Peacock's *Life* contains neither an index nor an itemised contents page, so that searching for a reference to a particular event is not easy. I have, nonetheless, looked through the most seemingly relevant chapters: VI, VIII, X, XII, XIV, and XV.

[19] Mary Somerville, *Personal Recollections* [n.1], 131. For Young's work on Egyptian hieroglyphs after 1814, see George Peacock, *Life of Thomas Young* [n.18], 258–344.

Before 1826, when the Royal Society published, in its *Philosophical Transactions*, her paper on the ability of the violet rays of the solar spectrum to induce magnetism in iron, Mary Somerville had no published works to her name. It is interesting to consider, therefore, how she had already acquired an established reputation as a scientist by that date. I believe that her early reputation came about because of the sociable character of this Grand Amateur community. The remarkable openness of the Edinburgh and London intellectual worlds, their lack of emphasis upon formal qualifications, and the prominence of the soirée, conversazione, dinner-party, and other informal sociable gatherings, made them places where talent could rise to prominence. It is true that this was also a very narrow world, restricted to no more than a few hundred people who, in addition to intelligence, enjoyed relatively comfortable economic circumstances—be they a modest £300 a year private income, or the revenues drawn from ancestral broad acres.[20] There was very little chance, therefore, of working people rising into it—not just because of their poverty, but also, sadly, because of the lack of opportunity available to them to acquire the essential education and leisure to pursue extensive and often abstract intellectual enquiries.

This Grand Amateur intellectual world, moreover, was not exclusively scientific. Its members would not have considered themselves to be 'scientists' in the modern sense, so much as 'literary and philosophical' people, their soirées including as they did lawyers, reforming politicians, poets, philosophers, economists and essayists. Jane and Thomas Carlyle, Sir Robert Peel, Lord Henry Brougham, Michael Faraday, Sir Charles Lyell and Lord Thomas Macaulay were all part of it. On one occasion, when Peel chanced to call in upon Faraday at the Royal Institution, he was shown, in the laboratory, Faraday's *experimentum crucis* whereby circular motion was generated from electromagnetic induction to produce the dynamo. The politician, who was especially interested in practical inventions, asked the physicist what possible use it could have. Faraday replied 'I know not, but I wager that one day your government will tax it'.[21] Faraday's apparatus, in fact, was the first electric motor!

In this world, therefore, it is perhaps easier to understand the fame of an unpublished scientist than it would be today. Quite simply, Mary Somerville became famous via her letters, her conversation, and by the fact that everybody in intellectual London knew of the extraordinary woman who had mastered the most abstruse mathematics of the age, and had acquired from her studies a sophisticated grasp of how physical science worked (Fig. 2.2).

It was mentioned above that in July 1817 the Somervilles crossed the Channel on their first continental visit together. We must also remember that at this time there was a whole generation of British people who had never been abroad, for between 1792 and 1815 the once familiar route of the Grand Tour had been blocked due to the French Revolutionary and Napoleonic Wars, except for a few months

---

[20] A. Chapman, *The Victorian Amateur Astronomer. Independent Astronomical Research in Britain 1820–1920* (Praxis-Wiley, Chichester and New York, 1998), 3–11. For Mary Somerville's £300 Civil List Pension, see Elizabeth Patterson, *Mary Somerville and the Cultivation of Science* [n.3], 153–155.

[21] L. Pearce Williams, *Michael Faraday* (Chapman and Hall, London, 1956), 196.

Fig. 2.2 A lecture on chemistry being delivered at the Surrey Institution, London, 1814. Note the large number of ladies in the audience. In its day, the Surrey Institution (which no longer survives) was a rival to the Royal Institution, which drew a similar proportion of ladies. Astronomy and chemistry were very popular subjects (Author's collection.)

during the fragile peace of the Treaty of Amiens. Only a tiny handful of British intellectuals—such as Sir Humphrey and Lady Jane Davy, accompanied as they had been by the young Michael Faraday in 1813—had been permitted to travel abroad. After Waterloo in June 1815, however, the English upper classes flooded onto the continent once again, and 2 years later the Somervilles joined the throng. Yet it was plain that even by this early date, some 8 years before she published anything, Mary Somerville arrived in Paris not as a tourist, but as a woman of note.[22]

Her reputation had clearly spread to Paris by letter and by word of mouth. Very significantly, however, she had met in London the French physicist Jean Baptiste Biot, who was in England working on a geophysical survey, and it was his wife in Paris who became the principal introducer of the Somervilles to the French *savants*. Madame Françoise Gabrielle Brisson Biot was an educated woman who had translated a German scientific text into French, though it was, as Mary Somerville recorded, 'published under the name of her husband'.[23] In Paris she also was

---

[22] Mary Somerville, *Personal Recollections* [n.1], 108–121. The original Journal of her 1817–1818 continental travels is in a small green notebook commencing '17th July. With a fair wind we embarked at Dover 10 min before 12 in the King George Packet...': Bodleian Library, Somerville Papers, Dep. c. 355 MSAU-1 Book No. 2.

[23] Mary Somerville, *Personal Recollections* [n.1], 110.

entertained by Dominique Arago and his wife at the Paris Observatory, and was shown 'all the instruments in the minutest detail, which was highly interesting at the time, and proved more useful to me than I was aware of'. We must remember in this respect that up to this point Mary's astronomical knowledge had been derived almost entirely from books. Though she knew in theory how astronomical observations were made, it is probable that before 1817 she had never been inside a major observatory, with the exception of her short visit to that of Sir William Herschel at Slough the year before. No doubt this is why Dominique Arago's tour of the Paris Observatory, and his detailed description of how to use the instruments, later 'proved more useful' than she was aware of at the time. Such a knowledge of practical scientific procedures would have been essential in framing in her own mind the background to her subsequent *On the Mechanism of the Heavens* and *On the Connexion of the Physical Sciences*.

Through Arago, Mary Somerville was introduced to Laplace, the most illustrious of all the French men of science, and the two developed an enduring intellectual friendship which would lead to her producing a great synopsis of and developing commentary on his work in *On the Mechanism of the Heavens*. Alexis Bouvard, Louis Poinsot and the palaeontologist Baron Georges Cuvier also entertained her.

Although England and France had fought bitterly over 22 years, the *savants* of both countries recognised and esteemed each other. English astronomers and mathematicians admired Laplace, Lagrange, Bouvard, and others, while English geologists like William Buckland and Sir Roderick Murchison (who had fought in the British army during the Peninsula campaign against Bonaparte) admired Cuvier. The French, in their turn, admired the quality of the observational data being produced by the Greenwich Observatory, and in particular held the late Revd Dr. James Bradley, Sir Isaac Newton, and the chemists John Dalton and Sir Humphry Davy, in the profoundest respect.

French science, however, was organised very differently from that of Britain, for both under the *ancien regime* and under Bonaparte a highly centralised, state-funded and patronised, and almost entirely Paris-based system of working prevailed, that was the antithesis of the diverse culture of the British Grand Amateurs. Indeed, so Paris-centred was French intellectual life that Cuvier informed Mary that when he had been sent to inspect the schools of Bordeaux and Marseilles, 'he found very few scholars who could perform simple calculation in arithmetic',[24] while many were ignorant of 'science, history… literature', and even the writings of the French philosophers. Indeed, 'Cuvier said such a circumstance constituted one of the striking differences between France and England; for in France science [was] highly cultivated but confined to the capital'.

From Paris, the Somervilles travelled to Geneva, to where Mary found that her reputation had also spread, and where she met Jane Marcet and her husband Alexander John Gaspard Marcet. It had been Jane Marcet's popular books on chemistry which had first stimulated Michael Faraday's interest in science around

---

[24] Mary Somerville, *Personal Recollections* [n.1], 112–113.

1809.[25] From Switzerland, they travelled into Italy—a country with which she was to fall in love, to which she would return, and in which, 54 years later, she would die. And here again, she was to find that her reputation had gone before her. She was, in spite of being an unambiguous Scotch Protestant, graciously received by His Holiness Pope Pius VII, 'a handsome, gentlemanly, and amiable old man' who blessed her[26]; after which, Mary and William travelled to Naples, where they would make a dramatic and dangerous ascent of the cone of Mount Vesuvius.

The Somerville's continental progress came to an end when William returned home to take up his newly-Gazetted plum post as Physician to Chelsea Hospital in 1819. Mary's reputation continued to grow in France, Germany, Switzerland, and Italy, of course, and on their next visit in 1832, in the wake of the publication of her Laplacian *On the Mechanism of the Heavens,* the Parisian newspapers hailed her coming as that of an intellectual celebrity. By 1832, with two major publications to her credit, and with clearly more to come, Mary Somerville's former verbal reputation had been transformed into that of a major published *savant.*

By the time of her visit to Paris in 1832, Laplace was dead, Alexis Bouvard was Director of the Observatory, and the fiercely Republican Dominique Arago was involved in politics after the Revolution of 1830.[27] Even so, Mary was now feted by everyone, and while she very clearly enjoyed the company in which she moved, she nonetheless took a particular dislike to the physiologist François Magendie, finding him coarse and uncouth. One suspects that she had, in any case, a prior aversion to Magendie because of the notorious cruelty of his vivisection experiments, contrasting him with Sir Charles Bell of Edinburgh who 'made one of the greatest physiological discoveries of the age without torturing animals'.[28] Both Bell and Magendie had discovered the ways in which nerve fibres are grouped and transmit signals, though Bell's work slightly preceded that of the Frenchman.

Indeed, Mary's social round in Paris in 1832 was remarkable in its range, as not only scientists but public figures like General Marie Joseph Lafayette and the visiting American novelist James Fennimore Cooper and his wife were keen to enjoy her company. And by the time that Mary and William Somerville returned to England, her international reputation as a woman of science was assured.

---

[25] *Jane Marcet, Conversations on Chemistry, in which the Elements of that Science are familiarly explained and illustrated by Experiments,* 2 vols. (1806). L. Pearce Williams, *Michael Faraday* [n.21], 19, 20.

[26] Mary Somerville, *Personal Recollections* [n.1], 121–122.

[27] Mary Somerville, *Personal Recollections* [n.1], 185–186. *Elizabeth Patterson, Mary Somerville and the Cultivation of Science* [n.3], 100–108.

[28] Mary Somerville, *Personal Recollections* [n.1], 192–193.

References

1. Lubbock, C. A. (1993). *The Herschel Chronicle. The life-story of William Herschel and his sister Caroline Herschel* (pp. 73–74). Cambridge: Cambridge University Press.
2. Nasmyth, J. (1889). James Nasmyth, engineer: An autobiography. In S. Smiles (Ed.) (pp. 18–62). London: Smith, Elder & Co.
3. Kemp, M. (1970). Alexander Nasmyth and the style of graphic eloquence. In *The Connoisseur* (93–99).

Chapter 3
The Domain of Nature: Astronomy, Optics and Geology

While there were major differences of national style, funding and career paths between the scientific communities of England, France and the other Continental countries, the scientists themselves shared the same broad intellectual concerns. It is true that while some countries had their especial distinctions—France for pure mathematics and physiology, Germany for organic chemistry and manufacturing optics, and England for experimental physics and observational astronomy—all of these scientists acknowledged the same concepts of nature, and recognised parallel standards of excellence. In her writings between 1825 and 1869, Mary Somerville was to explore this world of ideas, showing herself to be an ingenious experimentalist on the one hand, and a brilliant surveyor, interpreter and high-level communicator of contemporary science on the other.

A proper assessment of Mary Somerville and her career requires a discussion of the sciences as they were understood in her day, and it is important to identify the intellectual concerns of early-nineteenth-century European science and to place them into context. And since her active scientific career, from self-taught student to her death, spanned over 70 years, one must realise that science itself progressed enormously over that time. But while physical science had so advanced between 1800 and 1872, it had not developed 'beyond recognition', for the conceptual world of the science of the 1870s was logically and experimentally related to that of the Georgian age. Yet the sheer volume and extent of what had taken place, in both pure and applied science, during this period had been truly prodigious. Electricity had gone from being a novel natural force to an established telegraphic and carbon-arc lighting technology; astronomy had added new planets to the Solar System and identified the presence of standard laboratory chemicals in the Sun and stars; geology had opened up a wholly new concept of Earth history, from which Charles Darwin was to develop evolutionary theory; while anaesthesia, organic chemistry and the discovery of bacteria were to change William Somerville's profession truly beyond recognition.

But as we have seen, Mary Somerville was not a life-scientist but a physical scientists by instinct (although she had a great love of living things, especially birds), and her published research papers after 1825 and the subsequent editions of her books

© The Author(s) 2015
A. Chapman, *Mary Somerville and the World of Science*,
SpringerBriefs in History of Science and Technology,
DOI 10.1007/978-3-319-09399-4_3

after 1831 make it possible to identify where the priorities of science lay during her creative lifetime: in astronomy, the nature of light, geology, and physical geography.

Astronomy

Closest to Mary Somerville's heart was astronomy which since the days of Galileo and Kepler two centuries earlier had been, and still remained, the most advanced of the sciences. Astronomy's head start as the physical science *par excellence* came from the way in which exact instrumental observations of the heavens had been interpreted and given predictive power through mathematics. As instrumental observations improved, so did the mathematical techniques, via Kepler, Newton, Halley, Lagrange and Laplace, so that by 1830 astronomy had long been acknowledged as the role model for the other sciences. For while as yet the botanist or zoologist could only compare shapes, colours, and habitats, and the physician was eternally bamboozled by the nature of infection, the astronomer could predict the exact movements of Jupiter for over a 100 years hence.

Yet as astronomy raced ahead between 1800 and 1872, it is possible to identify particular questions about the nature of the Universe that its researchers were asking which dominated the discipline intellectually, and were discussed in Mary Somerville's books.

One of the most important of these questions pertained to binary and triple stars. It had been William Herschel after 1782, following the suggestion of the Revd John Mitchel, who had discovered the first sets of stars that appeared to be so close together that he suspected that they might be gravitationally connected.[1] If stars that were so remote in the depths of space as to appear from Earth only as dim specks could be shown to be gravitationally connected in a pair, then this opened up one of the profoundest truths ever discovered by science: that the same laws which held the Solar System together also governed the stellar Universe. While it had been presumed that this would be so—it being implicit, after all, in Newton's 1687 universal laws of gravitation—a concrete physical demonstration of such universality would indeed be a *coup*. Such a demonstration would prove that one law held the whole of the physical creation together, from the falling of sand grains to the mutual attraction of pairs of stars. Moreover, that law could be expressed mathematically and comprehended by the human intellect.

If Herschel's stars really were binaries, then the stars constituting the pairs should rotate around each other in elliptical orbits over a period of time, be it decades or centuries. If their mutual movements could be measured with powerful telescopes, then their orbits could be computed, and the masses or sizes of the individual stars and the precise geometry of their orbits could be calculated. Between 1802 and 1830, astronomers in observatories across Britain and continental Europe began to discover and measure dozens of such paired stars. By 1830, Félix Savary in France had computed the gravitational elements of the star ξ [xi] Ursae Maioris, while in England Sir John Herschel (Sir William's son) successfully

---

[1] See Ref. [1–3].
[2] See Ref. [4].

determined the gravitational characteristics of further double stars.[2] This was a formidable achievement, showing the power of the human intellect to use observation combined with mathematics to plumb the profoundest secrets of nature.

Double stars, their careful measurement, and the computation of their orbits, figure prominently in Mary Somerville's correspondence with Sir John Herschel, especially during the 1830s. In June 1831, for instance, they were discussing William Rutter Dawes' (of Ormskirk) 'excellently' measured positions of important binary stars such as ξ [xi] Ursae Maioris and 70 Ophiuchi, along with the work of Savary in France and of Johann Encke in Germany.[3] Mary described this work, and that of other double-star astronomers, in *On the Connexion of the Physical Sciences*.

It was perhaps in this context that Mary Somerville later recalled her indebtedness to Arago for providing her with such precise instruction in how to use the measuring instruments in the Paris Observatory in 1817. For in order to thoroughly understand the mathematics of double-star astronomy, it was also essential to know how the critically accurate observations of the particular pairs of stars were made in the first place, to obtain that firm bedrock of reliable data upon which a mathematical investigation could be based.

Because of the social customs of the day, and because of the widespread belief that women were especially susceptible to chill night air—a belief still reiterated 60 years later, when Elizabeth Brown recommended solar study to aspiring female astronomers—women did not work in observatories, nor study the stars through telescopes during the long watches of the night.[4] Mary therefore needed some practical observational experience.

It is possible that her first mentor in practical observatory astronomy—at least in England—was the surgeon Sir James South, who maintained observatories first at Blackman Street, Southwark, and then at Campden Hill, Kensington. South was a leading double-star observer between 1816 and the mid-1830s, and Mary recorded a visit by herself and her husband to Sir James and Lady South at Campden Hill, where she 'learnt the method of observing, and sometimes made observations myself on the double and binary systems, which, worthless as they were [for exact scientific purposes], enabled me to describe better what others had done'. Such observations, along with those of planetary positions, made with fine refracting telescopes and delicate micrometers in South's observatory, gave her hands-on experience of data collection and analysis, for 'when I took the mean of several observations, it differed but little from that which Sir James had made; and there I learnt practically the importance of taking the mean of approximate quantities'. Double-star observing required great skill and constant practice to do it well.[5]

The dates of these visits to South are not recorded, but they no doubt fell between 1826, after he transferred his observatory from Southwark to the then

---

[3] Sir John F.W. Herschel to Mary Somerville, 6–11 June 1831, Royal Society, Herschel Papers, HS16 (345). See also John EW. Herschel, *Outlines of Astronomy*, 2nd edn. (1849), 564–578.

[4] See Ref. [5].

[5] See Ref. [6].

cleaner air of Kensington, and the late 1830s, when his increasing paranoia virtually excluded him from the active scientific community. In addition to Mary Somerville, Sir James South had given previous training in observatory practice to the young John Herschel and the even younger George Airy in the early and mid-1820s. South's growing rage and mental instability after about 1840, indeed, was a tragic loss to British Grand Amateur astronomy.

Mary Somerville was also invited to make observations at the private observatory of the eminent double-star Grand Amateur Admiral William Henry Smyth. Between October 1835 and March 1836, Smyth kept her informed of the motions of the two components of the double star γ Virginis. The secondary component of this star describes a highly elliptical orbit around the primary, and during 1835–1836 the secondary passed through periastron—its closest approach to the primary. In January and February 1836, the components were so close to each other that both Smyth's 5.9-inch Tulley refractor and Herschel's 18¼-inch reflector showed γ^1 and γ^2 Virginis as a single star.[6] This was the first time that such a phenomenon had been observed, and it provided further data from which calculations demonstrating the precise action of Newton's laws in stellar space could be drawn. And whereas South was difficult and cantankerous, Admiral Smyth was genial, fun-loving, and surrounded by friends. Mary also became friendly with Smyth's gifted daughters, especially Miss Annarella (Fig. 3.1).

In addition to the observation and computation of binary stars, another area of cutting-edge research in astronomy was the study of *star clusters* and *nebulae*. On superficial examination—either with the naked eye or with opera glasses or small telescopes—clusters and nebulae appeared as diffuse patches of light in the sky, covering a larger angular area yet much dimmer than normal stars. The intellectual puzzle, which lay at the heart of cosmology, was: what were they? What was their distance from the Sun, and how did they relate to the stars of the Milky Way? For inevitably, they posed the grand question of exactly how the Universe was structured (Fig. 3.2).

From the 1780s onwards William Herschel, using powerful reflecting telescopes, had begun to 'sweep' the sky for these objects, and by 1800 had found more than 2,000 of them. Were the clusters—such as the familiar Pleiades, or the dense starfield in the constellation of Hercules—zones of intense deep-space gravitational attraction, forever sucking surrounding stars into a dense, compressed locus of force? And were the nebulae—such as those glowing in Orion's Sword or Andromeda—vast 'congeries' of stars that were so massive and so dense that even in the most powerful telescopes no individual stars could be seen; only the

---

[6] William Henry Smyth, Bedford, to Mary Somerville, 3 October 1835 and 26 March 1836, reproduced in Mary Somerville, *Personal Recollections* [n. 5], 210–213. W.H. Smyth, *A Cycle of Celestial Objects,* vol.2, the Bedford Catalogue (London, 1844), 275. W.H. Smyth, *Aedes Hartwellianae, or Notices of the Manor and Mansion of Hartwell* (London, 1851), 312–342 for 'The story of γ Virginis'. I am grateful to R.A. Marriott for drawing my attention to Smyth's observations in the Bedford Catalogue, and for providing me with details concerning the orbit of γ Virginis.

Fig. 3.1 The equatorially mounted 5.9-inch refractor by Tulley of London, owned by Admiral W.H. Smyth of Bedford. With this telescope, Smyth made crucial double-star observations, about which he corresponded with Mary Somerville in 1835-6 (Here it is prepared for observations of the transit of Venus in 1874.) (Royal Astronomical Society, Add. MS 93 No. 112.)

combined glow of many millions of stars? And if this was so, how vast must the Universe be, and how universal the forces that bound it all together! And how potent a reminder they were of the awesome power of the God who had made them all, and had given mankind the faculties necessary to comprehend His Creation by means of observation and mathematical analysis![7]

---

[7]   See Ref. [7].

Fig. 3.2 Herschel House, Slough, visited by Mary and William Somerville in 1816 (R.S. Ball, *Great Astronomers*.)

In an age especially attuned to the sublimity of mountains and sunsets, one can easily see how problems in cosmology struck the Georgians and early Victorians to their depths, and posed innumerable questions in theology and philosophy. But to see the star clusters and nebulae to the best advantage, one needed not the refracting telescopes used for binary work (where critical measurement was foremost), but reflecting telescopes, in which the image was formed not by a lens, but by a large 'speculum' metal mirror the surface of which was figured into an exact parabolic curve. Such mirrors, of 18, 24 and 48 inches diameter, could collect enormous quantities of light, so that dim and diffuse objects could be focused and studied.[8]

These big reflecting telescopes had been developed in the late eighteenth century by William Herschel at his private observatory in Slough, which William and Mary Somerville had visited in 1816 (the year in which Herschel received his knighthood); and following Herschel's death in 1822, the great cosmological project of fathoming the 'length, depth, breadth and profundity' of the Universe was carried on by his son, John.

John Herschel (knighted in 1831) became Mary Somerville's closest astronomical friend and adviser. He read through the manuscript draft of all her books, adding comments and advice (which still exist), and exchanged numerous letters

---

[8] See Ref. [8].

with her between 1830 and 1869. Her *Personal Recollections* record visits to the Herschels at Slough, and in particular, viewing nebulae and star clusters through John Herschel's famous reflecting telescope of 20 feet focus with its 18¼-inch mirror;[9] and though she could not expect to see the same amount of detail in these objects as could Herschel—with his vision honed by relentless nightly observation —she clearly felt, as a scientific author, that it was her business to see them at first hand, and to be familiar with the techniques of observation.

When Herschel was reducing his southern hemisphere observations for publication in the mid-1840s, moreover, he often mentioned, in his letters to Mary, particular strangely-shaped nebulae, such as 30 Doradus, which in his powerful telescope in South Africa had resembled 'a true lovers' knot'.[10] The thousands of nebulae discovered by Sir William and Sir John Herschel in two hemispheres seemed to present a bewildering array of shapes and light intensities, and only the vaguest attempts to explain or understand them were possible at that time. By November 1843, however, news of the celestial wonders that were visible through Lord Rosse's 36-inch-aperture telescope in Ireland had reached Rome, and Mary wrote to Herschel to ask for confirmation: 'I do not know what to think [?] of Lord Rosse's telescope if the accounts are not exaggerated it must be wonderful indeed but I only know it from public report' (Fig. 3.3).[11]

On 11 November 1843, moreover, Mary had written directly to the Third Earl of Rosse.[12] His reply on 12 June 1844 made crystal clear the whole significance of contemporary cosmological research. Lord Rosse was of the opinion that all nebulae were really massive star clusters, and that 'it is impossible not to feel some expectation that with sufficient optical power the nebulae would all be reduced to clusters'.[13] And less than a year later, when his giant 72-inch telescope with its 52-foot tube, superseding the 36-inch, came into use at Birr Casle, Rosse felt fully justified in this belief when he resolved one of Herschel's round nebulae into a spiral structure which he immortalised as the 'Whirlpool'. All of this work in Britain was undertaken by astronomers in the independent Grand Amateur tradition; for as Mary wrote (presumably during the 1860s): 'There are twenty-six private observatories in Great Britain and Ireland, furnished with first-rate instruments, with which the most important discoveries have been made'.[14]

---

[9] Mary Somerville, *Personal Recollections* [n. 5], 105, 134.

[10] John F.W. Herschel to Mary Somerville, 17 March 1844. Royal Society MS, Herschel Papers HS16 (348).

[11] Mary Somerville to John F.W. Herschel, Rome, 12 November 1843, Royal Society MS, Herschel Papers HS16 [347].

[12] Mary Somerville to Lord Rosse, Rome, 11 November 1843, Rosse Archives, Birr Castle [Ireland], K. 17:16.

[13] Lord Rosse to Mary Somerville, 12 June 1844, Rosse Archives, Birr Castle, K. 17 Additional [Birr, Ireland]. Reproduced in Mary Somerville, *Personal Recollections* [n. 5], 215.

[14] For Lord Rosse's discoveries, see Patrick Moore, *The Astronomy of Birr Castle* (Birr 1981). For the twenty-six private observatories, see Mary Somerville, *Personal Recollections* [n. 5], 270. Indeed, these twenty-six would include only the *major* private research observatories, for in its

Fig. 3.3 The globular star cluster Messier 13, in the constellation Hercules. Astronomers were fascinated by this and similar dense star clusters, and correctly surmised that their formation was due to dense gravitational attraction. *Upper* 'resolved into stars' (J.F.W. Herschel, *Outlines of Astronomy*); *lower* a recent CCD image by N. Szymanek and R. Dalby

(Footnote 14 continued)
census of British private observatories in 1866, the periodical *Astronomical Register* 4 (1866), 21, 91, lists forty-eight significant observatories. See also A. Chapman, *The Victorian Amateur Astronomer* [n. 8], 228.

Fig. 3.4 Messier 51, the Whirlpool Nebula, in the constellation Canes Venatici. This 'nebula' (now known to be a galaxy) appeared as a misty patch in earlier telescopes, but in 1845 Lord Rosse's 72-inch telescope revealed its complex structure, suggesting that it was a massive 'whirlpool' of stars in deep space. *Upper* Lord Rosse's first drawing (*Philosophical Transactions of the Royal Society*, 1850); *lower* a recent photograph by R.W. Arbour

However, as the Grand Amateur William Huggins and the astronomers who began to analyse starlight with the first spectroscope came to discover in the mid-1860s, much of the diffuse matter in the nebulae did not consist of simple stellar aggregations, but was made up of clouds of glowing gas and dust (Fig. 3.4).[15]

---

[15] Hoskin, *Cambridge History of Astronomy* [n. 1], 292–293. For a history of astronomical spectroscopy, sec J.B. Hearnshaw, *The Analysis of Starlight. One Hundred and Fifty Years of Astronomical Spectroscopy* (Cambridge University Press, 1.986), 71.

Yet long before 1864—when the spectroscope confirmed the presence of gas-eous material in space—astronomers suspected its existence. Indeed, Sir William Herschel had discussed the possible existence of 'luminous fluid' and glowing 'chevalures' in space since 1791. Not only did some star clusters seem to contain glowing diffuse matter as well as actual stars, but certain objects much closer to home and within the Solar System hinted at the presence of gas, or 'phosphorescent vapour', and what Sir John Herschel later called 'particulate matter', or space dust (Fig. 3.5).[16]

Mary Somerville devoted several pages of *On the Connexion of the Physical Sciences* to this topic, in her discussion of comets, meteors, the aurora borealis, the zodiacal light and other phenomena which suggested the presence of a glowing light without any apparent solid substance. Was there some primary, tenuous matter which pervaded the entire Universe, forming cometary tails and meteors, and out of which the Solar System had condensed under the force of gravity, as Laplace had proposed in his nebular hypothesis of 1796? And did this matter also condense to form diffuse objects like the Orion Nebula, and finally produce individual burning stars, in the same way that a cloud of steam condenses into millions of separate water droplets?

One strong piece of evidence suggesting the presence of this primary matter throughout space had been the comet of 1770. For more than a century, astronomers had believed comets to be massive, planet-like bodies; but the comet of 1770 produced several surprises. It was the first 'near-Earth' body on scientific record, 'grazing' the Earth—as Mary reminds us in *On the Connexion of the Physical Sciences*—at only six times the distance of the Moon—in astronomical terms, to within a cat's whisker! Yet when Jean-Baptiste Joseph Delambre in France ana-lysed the superbly accurate Greenwich Observatory records for 1770 to determine whether this comet had caused fluctuations in the length of the terrestrial day, the tides, and other minutely-recorded phenomena, he found not the slightest hint of disturbance.[17] This comet, and no doubt all other comets, was obviously a light-weight body made of a small nucleus and a wisp-like tail, and even at this close approach had no discernible gravitational effect upon the Earth. Were comets and meteors, therefore, nothing more than local agglutinations of the same glowing interstellar dust that also composed the nebulae?

Like most other scientists of her time, Mary Somerville felt that this supposed tenuous matter scattered throughout space must be somehow connected to the 'ether'.[18] This ether was invented to explain the transmission of light through space, as well as account for the anomalous movements of awkward planets such as Mercury and Uranus, for which calculations based upon actual observations of

---

[16] John F.W. Herschel, 'Observations of Nebulae and Clusters of stars made at Slough, with a Twenty-feet Reflecting Telescope, between 1825 and 1833', *Philosophical Transactions of the Royal Society* 123 (1833), 501. John Herschel, *Outlines of Astronomy* [n. 3], 598–600.

[17] See Ref. [9].

[18] Mary Somerville, *On the Connexion of the Physical Sciences* [n. 17], 356.

Fig. 3.5 To eighteenth- and nineteenth-century astronomers, comets were objects of great fascination because of their peculiar elongated orbits. *Upper* A contemporary drawing of the comet of 1819 (J.F.W. Herschel, *Outlines of Astronomy*); *lower* comet Hale-Bopp, photographed by S. Parkinson in 1997

these heavenly bodies never quite matched the predictions laid down by Newtonian gravitation. Uranus's erratic motions were spectacularly accounted for by the discovery of the planet Neptune in 1846—although Mercury's eccentricities were to need Einstein's theory of relativity to provide a true explanation only after 1916.

The nineteenth century was also a period of great progress in Solar System astronomy, with figures like William Rutter Dawes, William Lassell (who, like Lord Rosse, was building large reflecting telescopes at his own expense), and William Cranch Bond at the Harvard Observatory—all of whom were discovering new planetary satellites and learning about the physical stability of Saturn's rings. Needless to say, these discoveries figured prominently in Mary's correspondence with Herschel[19]; and when solar physics and astrophysics came into existence in the late 1850s, and scientists first began to investigate the physical processes by which the Sun and stars burned, these matters also featured conspicuously in the letters that passed between Italy and the Herschel mansion in Kent.[20]

Yet whether one was measuring binary stars, attempting to resolve nebulae into clear physical systems, discovering new planetary satellites and solar flares, or searching for evidence of an all-pervading physical medium, the wider astronomical enterprise was the same insofar as it aspired to tame the seemingly jumbled vastness of space, and reduce it to exact descriptions and precise mathematical expressions. This, after all, had been the intention behind Laplace's *Mécanique Céleste*, and was in turn to inspire and direct Mary Somerville's own thoughts and writings.

Optics

Like all of the physical scientists of the day, Mary Somerville was fascinated by the nature of light. Not only was it the agent by means of which all astronomical knowledge was derived, but the years 1800–1804 saw the discovery of parts of what would later come to be called the 'electromagnetic spectrum'. Much of this work was done in England, and in particular by three men who would later exert a formative influence on Mary's very concept of science.

In his attempt to develop filters through which he could directly study the solar surface through a large telescope (a practice which today would be considered foolhardy in the extreme), William Herschel noticed that filters of equal density yet differing colours imparted a greater or lesser sensation of heat to his eye, with red being the warmest. Herschel then proceeded to carry out laboratory experiments on radiant heat, and on 11 February 1800 found that when he passed a thermometer through a solar spectrum generated by a glass prism, the red light registered the greatest heat. Yet when he moved the thermometer beyond the red light, and into the dark, the temperature continued to rise, reaching a peak about one inch away from the red. He called this the region of 'invisible light'. Mary Somerville was to refer to the 'calorific' or hot rays of the spectrum. We now call this part of the spectrum the infrared.[21]

A year later, the German chemist Johann Wilhelm Ritter, who was conducting experiments into the effects of sunlight on silver chloride crystals, found that while the blue end of the spectrum had no heat properties, there was an area in the apparent dark, beyond the violet extremity of the visible spectrum, where silver chloride still turned black, just as it did in normal light. Once again, Mary

---

[19] John F.W. Herschel to Mary Somerville, 2 March 1851, Royal Society MS, Herschel Papers HS16 [356].

[20] John F.W. Herschel to Mary Somerville, 11 April 1865, Royal Society MS, Herschel Papers HS16 [372].

[21] Angus Armitage, *William Herschel* (T. Nelson & Sons, London, 1962), 56–59. Mary Somerville, *On the Connexion of the Physical Sciences* [n. 17], 226.

Somerville was later to become fascinated by, and conduct her own original experiments into, what she styled these 'chemical rays' of the Sun which we now call ultraviolet.

Then, in 1802 Dr. William Wollaston, who would become one of Mary Somerville's most respected early mentors in scientific research, made an ingenious modification of Sir Isaac Newton's 1666 experiment of breaking up the solar light by passing it through a pinhole and into a prism. Wollaston replaced the pinhole by a fine slit one twentieth of an inch in diameter, and focused the emerging spectrum with a lens. He was surprised to find the colours much better defined, and traversed by seven black lines. Wollaston mistook these lines for natural divisions between the colours, but in 1814 a Munich optician would take his experiment further and discover more than five hundred 'Fraunhofer lines' in the Sun's spectrum. Joseph Fraunhofer realised that Wollaston's black lines were much more than simple colour boundaries, although in 1814 no-one appreciated their true significance. After 1860, however, these black Fraunhofer lines would supply the key to the new science of spectroscopy, which Mary Somerville would discuss in detail in the subsequent editions of her *On the Connexion of the Physical Sciences*.[22]

Throughout the period 1800–1804, Wollaston's friend Dr. Thomas Young— another major influence on Mary Somerville—was conducting experiments in London that would establish the *wave*, as opposed to the Newtonian *particle*, theory of light. Since Newton's original prism experiments in 1666, and as enshrined in his *Opticks* (1704), light had been envisaged by scientists as being made up of 'corpuscles', or fast-moving bullet-like particles. Different colour sensations were produced by particles, moving at different speeds, hitting the eye. Yet even in the 1670s, scientists such as Robert Hooke and Christiaan Huygens had questioned the Newtonian 'corpuscular' model, and had explained peculiar optical phenomena— such as the colours produced by mother-of-pearl—in terms of light moving in waves. But Newton's prestige became such that their alternative explanations were effectively ruled out in the eighteenth century (Fig. 3.6).

Soon after 1800, however, Thomas Young once again began to investigate light, and discovered the phenomenon which came to be styled 'interference'. His experiment was classic in its stark simplicity, yet it was to change mankind's entire conception of the nature of light and provide concepts and a terminology which would later be applied to the radio and other bands of the electromagnetic spectrum. Mary Somerville was to describe it with her characteristic clarity in the *Connexion*. Light that has first passed through a piece of coloured glass goes on to pass through a pair of fine slits close together (let us say, cut by a scalpel into a piece of cardboard), and falls as two lines on a white wall. It will be noted that the resulting coloured bands of light are not of even intensity, but have dark parallel streaks running across them. Yet if one of the slits is closed, the dark streaks vanish, and a

---

[22] Owen Gingerich, 'Unlocking the chemical Secrets of the Cosmos', in Gingerich, *The Great Copernicus Chase* (Sky Publishing, Cambridge, Massachusetts; and Cambridge University Press, 1992), 170–176. Hearnshaw, *Analysis of Starlight* [n. 15], 20–30, 40–52.

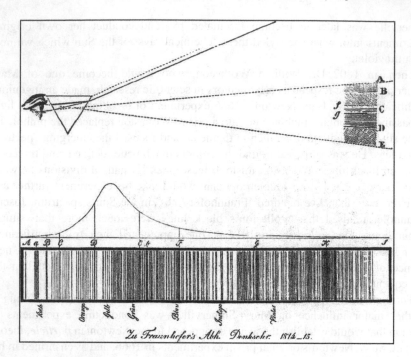

Fig. 3.6 *Upper* Francis Wollaston's first observation of *lines* in the solar spectrum (1802); *lower* Joseph Fraunhofer's first map of the solar spectrum (1815) (J. Norman Lockyer, *Stargazing*.)

pure colour streams through the other slit.[23] Mary then mentioned that, by way of amplification, Young stretched out a human hair in a beam of light passing through a pinhole. Parallel sets of dark bands once more appeared on the screen, though when one half of the image was masked with a card, the bands disappeared.

What Young had discovered was the optical phenomenon of *diffraction*, whereby sharp, fine edges, such as slits or hairs, interfere with the physical structure of a beam of light. Such interference, it was argued, could occur only if the light was travelling in tightly packed vibrating rays—different amplitudes or frequencies of which produced different colours—and could be split by passing them over a sharp edge. By analysing the exact geometrical relationship between the slits or hairs and the dark interference bands on the projection screen, she reported that the amplitude or size of the individual rays of light could be measured.

What really fascinated Mary Somerville about the spectrum, however, and which formed the subject of her three published research papers, were the 'rays, known by their chemical action, [which] exist in the dark space beyond the extreme violet', and were, no doubt, 'modifications of the same agent which produces the sensation of light'.[24] In 1836, when she wrote these words, Mary had no actual proof that

---

[23] Mary Somerville, *On the Connexion of the Physical Sciences* [n. 17], 186–188.

[24] Mary Somerville, *On the Connexion of the Physical Sciences* [n. 17], 225–226.

these rays, along with their red 'calorific' counterparts, were part of the same natural energy band, though in reality she was correct.

It was the blue end of the spectrum which became the subject of Mary Somerville's first publication in the *Philosophical Transactions of the Royal Society* in 1826. Hearing that Professor Morichini in Rome had supposedly magnetised small pieces of iron by exposing them in the violet rays of the solar spectrum, and that others had failed to replicate his results, she addressed herself to the task over the clear days of August 1825. She fixed a glass prism into a window shutter, and using a 'very large lens'—loaned to her by Wollaston—to focus the violet rays, exposed carefully demagnetised needles and slips of old watch spring. She claimed that after 2 h of exposure, preferably some time between 10 am and 1 pm, her needles exhibited a clear magnetic charge when placed near a compass.[25]

As a good experimentalist, she began to vary the parameters of the experiment, placing the needles in or out of the local magnetic meridian, and found that indigo light also imparted a magnetic charge. She then went on to expose metal slips for six hours underneath slabs of glass coloured blue with cobalt, and again reported a magnetic charge; and she also tried exposing pieces of paper treated with silver chloride in the rays and noting the ways in which they darkened. 'I am induced to believe', she concluded, 'that the more refrangible [violet] rays of the solar spectrum have a magnetic influence even in this country' where those rays strike the Earth much more obliquely than they do in Rome. Oddly enough, however, this magnetising effect is neither understood nor really agreed upon even by physicists today.

At the end of March 2001, Sarah Parkin—a research student in the history of science at University College, Oxford—attempted to replicate these magnetising experiments. She built an apparatus similar to that described by Mary Somerville in 1826, wherein a prism was mounted within an aperture in a black card, so as to produce a spectrum. For 3 h around noon, a steel needle was exposed in the violet rays, but no magnetic charge was later detected in the needle. Sarah Parkin then repeated the experiment with a perspex prism—which has dispersive powers slightly different from those of glass—but once again, no magnetism was detected. Finally, a steel needle was exposed in the beam of an ultraviolet lamp (to determine whether it was the invisible ultraviolet as opposed to the violet light that was inducing magnetism); but once more, no magnetism could subsequently be discerned in the needle. As Sarah Parkin realised, the low March Sun may not have possessed the magnetism-inducing power—a factor of which Mary Somerville herself had been all too aware when working in northern Europe, although the experiment also failed when the needle was exposed to an electrically generated optical source. No-one has been able to replicate Mary Somerville's 1826 experiments on inducing magnetism by exposing steel to a spectrum source, and we do

---

[25] Mary Somerville, 'On the magnetizing power of the more refrangible solar rays', communicated by W. Somerville, M.D., F.R.S., *Philosophical Transactions of the Royal Society*, 116, 2 (1826), 132–139.

not know—her grasp of scientific technique being what it was—how she obtained her results.[26]

In 1836, Mary corresponded with Dominique Arago in Paris, giving the results of her experiments on the Sun's 'chemical rays' in which she had exposed silver chloride-treated paper beneath variously coloured slabs of glass, mica, and gem stones such as emerald and garnet, tourmaline and rock salt. She found that rock salt, indeed, transmitted the greatest number of chemical rays. Arago published her results in French in the new physics journal *Comptes Rendus*.[27]

By 1836, of course, Mary Somerville had two major books behind her and was celebrated in England and abroad. None other than the eminent Michael Faraday, at the Royal Institution, prepared the silver chloride for these experiments with his own hands, and advised her how to apply it to the strips of paper to optimise their light sensitivity, and how to 'keep it for a long time in good condition'.[28]

It was probably around this time, moreover, that Mary Somerville became acquainted with Mary Griffiths of New York. In December of what was probably 1838, Mary Griffiths sent Mary Somerville a 'little volume' which she had written, entitled *Discoveries in Light and Vision*, and which seemed to relate optical experiments. It appears from the letter that Mary Somerville had already sent a small book to Mary Griffiths in New York via 'Miss Martineau', who no doubt was the writer Harriet, who had spent 1834 to 1836 in America (and to whom Mary Somerville had supplied a letter of introduction to James Fennimore Cooper) . I have, however, been unable to find out any more about Mary Griffiths or the kind of researches she was pursuing. One can gauge Mary Somerville's international standing by this date from the fact that Mary Griffiths' letter from America was simply addressed 'To Mrs. Mary Somerville, London'.[29]

In June 1845, while living in Rome, Mary returned to her work on the solar spectrum, opening a manuscript note-book entitled 'Experiments on light Rome 1845'. This little volume, now preserved with the other Somerville manuscripts in the Bodleian Library, Oxford, contains her optical experiments performed down to

---

[26] Sarah Parkin, 'Mary Somerville (1780–1872): Her Correspondence and Work on Chemistry', Oxford University Chemistry Part II Thesis, 2001, 54–60. Copy in History Faculty Library, Oxford. A. Chapman was Sarah Parkin's thesis supervisor.

[27] See Ref. [10].

[28] Michael Faraday to Mary Somerville, 12 October 1835, in *The Correspondence of Michael Faraday*, vol. 2, ed. Frank A.J.L. James (Institute of Electrical Engineers, London, 1993), letter 821.

[29] Mary Griffiths (New York) to Mary Somerville, 12 December 1838 (year unclear from script; could also be 1836, but Elizabeth Patterson in her 'Handlist' to the Somerville Papers in the Bodleian Library reads the date as 1838): Bodleian Library, Somerville Papers, Dep. c. 370, MGS-2/MSE-1. I have not been able to find Mary Griffiths in the standard dictionaries of women scientists, though her *Discoveries in Light and Vision, with a Short Memoir containing discoveries of the mental faculties* (C. & G. Carvil, New York, 1836), 300 pp. with 3 plates, is listed anonymously by title in the *National Union Catalog, Pre-1956 Imprints*, vols. 144 and 218 (Mansell 1971), with a named reference 'Mary Griffith' to the anonymous title entry. I have not been able to trace a copy of this book in England.

October 1851.[30] What she was trying to do in these experiment was to ascertain the effects of different spectral colours on vegetable preparations such as beetroot juice, nasturtium juice, and other flower-petal extracts. Marked onto the pages, she denoted the exact boundaries of the spectrum, and the 'heat spots' produced by a 7½-inch-focus flint-glass lens. Still preserved in the notebook are several slips of paper, a couple of inches long, stained blue, brown, and so on, depending on the plant dye employed.

The results of these experiments were sent to Sir John Herschel, who promptly had them published in the *Philosophical Transactions of the Royal Society* in 1846. As in her two previous experimental papers of 1826 and 1836, Mary was not in pursuit of some grand conclusion, but reporting meticulously conducted researches into discrete pieces of optical phenomena: why the red 'calorific' rays of the spectrum seemed to have no effect on either silver chloride or vegetable juices, yet 'the most refrangible rays from the green to the end of the lavender darken some [vegetable] substances and bleach others'.[31] Alas, in 1845 not enough was known about the action of high-energy photons of light on complex organic molecules to go beyond careful description. Yet it is upon such descriptive researches into and classification of phenomena that true experimental science, as opposed to speculation, ultimately stands.

One problem which Mary Somerville and all other researchers into the chemical, thermal and magnetic effects of the solar spectrum must have experienced was that of keeping the image in one place as the Sun moved, especially if exposures were to last for several hours. It was clearly a subject of discussion between Mary and Sir John Herschel, for on 2 November 1845, when she would have been in the midst of her experiments on vegetable juices, Herschel suggested and sketched a simple equatorially mounted 'heliotrope' or Sun-tracker that could be powered by 'any common wooden clock', to keep the solar spectrum in one place for hours on end. Indeed, Herschel, as always, encouraged her, urging on her spectrum experiments, and writing: 'I always suspected there is a world of wonders awaiting disclosure in the solar spectrum'.[32]

But not even Herschel could have imagined how portentous this remark was to be. The 1850s were to see the rapid development of spectroscopy—first as a piece of chemical and physical laboratory apparatus, then after 1859, when Count Robert Bunsen and Gustav Robert Kirchhoff at Heidelberg detected the first chemical substances in the Sun, as an astronomical instrument. And by 1865, Father Angelo Secchi in Italy and William Huggins and William Miller in England had turned the

---

[30] Mary Somerville, a notebook with blue covers, bearing the title page 'Experiments on light Rome 1845', Bodleian Library Somerville MS MSSW-13 (Dep. 354).

[31] Mary Somerville, 'On the Action of the Rays of the Spectrum on Vegetable Juices. Extract of a letter from Mrs. M. Somerville to Sir J.F.W. Herschel, Bart., dated Rome, September 20, 1845', *Philosophical Transactions of the Royal Society,* Part II for 1846 (1846), 111–119: 118.

[32] Sir John F.W. Herschel to Mary Somerville, 21 November 1845, Royal Society MS, Herschel Papers HS16 [352], Also Mary Somerville, *Personal Recollections* [n. 5], 278.

spectroscope into a breathtaking research tool, to establish the science of astrophysics.[33]

Indeed, in April 1865, the 73-year-old Sir John Herschel, still feeling frail from winter attacks of influenza and bronchitis, wrote to the now widowed 85-year-old Mary Somerville in Italy, to discuss these new sciences of spectroscopy and astrophysics. Knowing from lifelong experience how dim was the light from distant nebulae, Herschel said that he could not understand how Huggins could get enough of it into his spectroscope to enable a detailed chemical analysis, before going on to describe an ingenious experiment which he, Herschel, had devised some years earlier to test the sensitivity of a prism in analysing a weak light source. On this occasion, Herschel had noticed that a dead lobster (no doubt intended to be eaten) glowed in the dark. A scientist to his fingertips, Herschel had the idea of passing the phosphorescent light of the lobster through a prism, only to find that its spectrum 'seemed to me all of one colour [and] I could not distinguish any *tints*'.[34]

And it had been Sir John Herschel, after 1839, who had done much to research the optical and chemical basis of the early art of photography, discovering, amongst other things, the process of 'fixing' a photographic image with 'hypo' or sodium thiosulphate.[35] And while Mary Somerville's original 1834 edition of the *Connexion* naturally said nothing about photography, her post-1840 editions included it. She saw in the formation and development of the photographic image a wonderful concourse of phenomena in which the 'chemical rays' of the Sun combined with the rapidly advancing science of chemistry to create both an exact recording medium for science and a potential art form.

The wave theory of light, the 'calorific' and 'chemical' rays of the Sun, the true nature of Fraunhofer's (or Wollaston's) lines, and the effects of the solar colours and their extremities on chemical substances, fascinated the scientists of the nineteenth century. And what they all concluded about this new model of light was that there must be an 'ether' or tenuous medium suffusing the Universe through which it passed. For just as air conveys those vibrations that produce the sensation of sound in our eardrums, and water those energy waves that sink ships, so light must have a medium through which to send ripples to create the sensations of light and colours in our eyes. For this reason, as Mary Somerville made clear in her books, the 'undulatory' or wave theory of light seemed to presuppose a 'medium' that suffused space.[36]

Through much of the nineteenth century, physicists assumed that such a medium must exist—largely because the model of nature built up by Georgian and early Victorian science, and which Mary's books had done so much to establish in the

---

[33] Owen Gingerich, 'Unlocking the chemical secrets of the Cosmos' [n. 22], 170–176.

[34] J.F.W. Herschel to Mary Somerville, 11 April 1865, Royal Society MS, Herschel Papers, HS16 [372].

[35] The best scholarly treatment of Herschel's photographic researches is found in Larry Schaaf, *Out of the Shadows. Herschel, Talbot, and the Invention of Photography* (Yale University Press, New Haven and London, 1992).

[36] Mary Somerville, *On the Connexion of the Physical Sciences* [n. 17], 356.

cultural consciousness, was deeply mechanical. Indeed, her use of words like 'mechanism' and 'connexion' in the very titles of her own works conveys this image of how she, and scientists of her generation, believed that nature worked: as a great and complex piece of clockwork in which 'energy' (a word brought into scientific usage by Thomas Young) always needed a neutral agent through which to pass, in the same way that the teeth in a gear train convey the force of the spring.

Not until the end of the nineteenth century, and after Mary's death, did new experimental evidence undermine this 'celestial clock' view of nature. Albert Abraham Michelson's and Edward William Morley's famous experiment conducted in 1887 to detect and quantify the ether drew a disturbing blank.[37] And while James Clark Maxwell's famous equations of 1864 finally demonstrated the long-suspected relationship between light, electricity and magnetism, they were not dependent upon the ether theory. But it would not be until the works of Albert Einstein, which appeared between 1905 and 1916, that the ether would finally be made redundant, and the Universe shown to be a place vastly more complex, multilayered and diverse than scientists of Mary Somerville's generation could ever have imagined. Even so, the late Georgian discovery of the existence and effects of the infrared and ultraviolet extremities of the spectrum opened up the clear possibility that light was not a simple phenomenon. And without that, the subsequent discovery of radio waves, X-rays, gamma radiation and the rest of the electromagnetic spectrum would not have been possible.

Geology

One of the most rapidly developing, and in some respects most controversial, of the sciences in the early nineteenth century was geology. It was rapidly advancing as a result of the discoveries in stratigraphy, fossil geology and comparative anatomy by William Smith in England and Baron Georges Cuvier in France, besides others, whose researches supplied the techniques whereby a scientist could arrange ancient events into temporal sequences, which seemed to suggest that the structures of living things had become more complex over vast periods of pre-human time.[38] It was controversial because these discoveries opened up questions of time scales, the necessity for the extinction of creatures which seemed to have no parallels to those mentioned in the Bible, whether Noah's Flood had been only the last of many prehistoric deluges, and how the human race was related to the rest of creation.

Fossils (Latin *fossilis*—'dug out', 'dug up') had been familiar for centuries, and many scientists from the 1660s onwards had realised that they were the petrified remains of once-living creatures. But while figures such as Robert Hooke and Edmond Halley had come to realise before 1700 that the Earth's surface had probably been remodelled many times, most people still thought of fossils as the

---

[37] For a clear account of the Michelson-Morley experiment and what it achieved, see Ian Ridpath, ed., *Collins Encyclopedia of the Universe* (Harper Collins, London, 2001), 86–87.

[38] See Ref. [11].

entombed remains of the victims of Noah's Flood.[39] This explanation, ironically, was even applied to fossilised fishes!

By 1820, however, geologists had come to realise that the Earth was immeasurably older than Adam and Eve, and that those creatures which inhabited the modern world were but the latest of many changes of species. As William Buckland suggested, the stratigraphic and fossil record implied that a series of global devastations had taken place in the archaic past. Each of these devastations—or catastrophes—had not only wiped out all contemporary life, but had resulted in great beds of detritus accumulating on the bottoms of the oceans, into which the remains of the exterminated creatures became embedded, to become new fossil-bearing strata. By counting the strata in a rock face, moreover, one could calculate the number of catastrophes that had taken place, and not fail to be impressed by the progressive complexity of living things found in each successive layer.[40] Early strata, for instance, contained no fossils at all—only crystalline granites and such. Then, in the non-crystalline sedimentary rocks one found that the most primitive plants were followed by more complex plants; that primitive shellfish were succeeded by fishes, reptiles, and then all the diverse tribes of those great land-dwellers which Sir Richard Owen in 1841 would style, from the Greek, *dinosaurs*, or terrifying, great lizards. And amongst these remains, moreover, one found no horses, cows, dogs or humans. Indeed, no archaic human remains were found until the latter part of the nineteenth century, which only seemed to confirm that our planet had a vastly ancient pre-human past.

How, therefore, could this geological realm be squared with the Bible, which suggested, from a count of the generations of the Old Testament Patriarchs, that humans were created as recently as 4004 BC, when God populated the Garden of Eden?

An ingenious explanation was put forward by Mary Somerville's friend, the Revd Dr. William Buckland, in his inaugural lecture as Regius Reader in Geology at Oxford in 1819. Buckland proposed that as the Bible was intended to be a book of divine guidance for mankind—which was a unique creation, made in God's image—then it had not been necessary to worry the ancient Jews with details about primitive ferns, ammonites, or megatheriums, none of which possessed immortal souls. For while the Bible began, quite naturally, with the plain factual statement that 'In the beginning God created the Heaven and the Earth', the *Genesis* narrative

---

[39] For a detailed examination of Hooke's geological ideas, see Ellen Tan Drake, *Restless Genius: Robert Hooke and his Earthly Thoughts* (Oxford University Press, Oxford and New York, 1996). For Halley's geological ideas see A. Chapman, 'Edmond Halley's Use of Historical Evidence in the Advancement of Science' (Royal Society John Wilkins Prize Lecture in the History of Science), *Notes and Records of the Royal Society*, 48, 2 (1994), 167–191: 180. A good popular (but somewhat dated) history of early geology is Herbert Wendt, *Before the Deluge. The Story of Palaeontology*, transl. Richard and Clara Wilson (Victor Gollancz, London, 1968).

[40] William Buckland, *Geology and Mineralogy considered with reference to Natural Theology*, 2 vols. (London, 1836). This 'Bridgewater Treatise' by Buckland provided an excellent survey of the science of geology and its intellectual assumptions by 1836.

discreetly skipped over many millions of years until the chaos of the last great catastrophe was clearing away—'And the Earth was without form, and void; and darkness was upon the face of the deep.' Buckland proposed that the *Genesis* narrative resumed at Chap. 1, verse 2 because it was only after this last catastrophe that God created the Garden of Eden, which was populated by modern fauna and crowned with the conscious presence of the first man and woman.[41]

By simply inserting the immensity of geological time between the first two verses of *Genesis*, Buckland had made it possible to reconcile the Bible with geology. And as the rest of Biblical history from the Garden of Eden onwards involved human beings, and narrated our own spiritual ancestry, nothing else needed modification. Buckland even interpreted Noah's Flood as the last geological catastrophe that had left its mark on the surface features of the continents.[42]

In this way, Buckland did an extremely valuable service for geology, for in predominantly low-church Protestant evangelical Britain, most people regarded Scripture as the direct and perfect utterance of God. And while there was an instinctive prejudice against reinterpreting Scripture, most thinking people none-theless were willing to countenance that God might not have put *everything* into what Galileo in 1615 had called 'The Book of the Word', and left us to use our divinely bestowed intelligence to discover scientific matters for ourselves in 'The Book of the World', or nature.[43]

In 1830, however, the Scottish barrister–geologist Sir Charles Lyell proposed another model to explain the Earth's development. Instead of periodic catastrophes, Lyell argued that a series of endless minor changes in the Earth's surface, produced by 'causes now in operation', such as volcanic and earthquake activity, along with erosion, could form all of the continents and mountains given enough time. And likewise, as habitats changed, whole species of living creatures would become extinct while others would thrive. Because Lyell saw geological processes as acting gently, yet inexorably, his system was known as 'Uniformitarian', in contrast with Buckland's 'Catastrophism'.[44]

Yet both Buckland and Lyell firmly abominated the early evolutionary theories put forward by the Frenchman Jean-Baptiste Lamarck in his *Philosophie Zoologique* (1809). Instead of evolution, both Buckland and Lyell saw all living things as divinely designed 'special creations', and while existing species might become

---

[41] William Buckland, *Vindicae Geologicae* (Oxford University Press, 1820), 31–32. Also, Buckland, *Geology and Mineralogy* [n. 40], 18–33.

[42] William Buckland, *Reliquiae Diluvinae* (London, 1823). In this influential essay, Buckland interpreted fossil bone caves and other phenomena in terms of the Flood of Noah, and pre-Noachian floods. Also Nicholaas Rupke, *The Great Chain of History. William Buckland and the English School of Geology* (Clarendon Press, Oxford, 1983), esp. 180–266.

[43] Galileo Galilei, *Letter to Madame Christina of Lorraine Grand Duchess of Tuscany. Concerning the Use of Biblical Quotations in Matters of Science* (1615), in *Discoveries and Opinions of Galileo,* translated and introduced with notes by Stillman Drake (Doubleday Anchor, New York, 1957), 175–216.

[44] Charles Lyell, *Principles of Geology,* vol. I (London, 1830), for the 'Uniformitarian' theory.

extinct, only God could create new species, which were, by definition, unique and anatomically unchanging.

Even so, one can glimpse here the ingredients for the emergence of a radically new concept of the history of the Earth: vast, pre-Biblical time scales, extinction, the use of comparative anatomy to study extinct and living forms, and an increasingly scientific, rather than theological, explanation of the Earth's formative processes. It was only a matter of time, therefore, before some ingenious individual rearranged all of these scientific ingredients afresh, to produce a radical and frightening conclusion. And that individual was the Edinburgh publisher and amateur geologist, Robert Chambers.

If one thought of the Earth in Uniformitarian Lyellian terms, yet had a naturalistic rather than a divine model for the origin and proliferation of living species, the result was a theory of 'transmutation', or simple evolution. Indeed, Chambers argued in his anonymously published *Vestiges of the Natural History of Creation* (1844) that life could well have emerged out of a primeval soup of chemicals activated by a lightning bolt. And given enough aeons of time, and variations of solar radiation and other environmental factors, these proto-creatures could well mutate, and fill the world with life.[45]

Vestiges was an instant bestseller. It shocked and fascinated Victorian society, and it is not for nothing that subsequent historians came to see it as 'Darwin's lightning rod' in that it was the reading public's first serious brush with evolution, some 15 years before *On the Origin of Species*.

Gradually accustomed to enormous geological time spans and subtle re-readings of *Genesis* as intelligent early Victorians were becoming, it would still be erroneous to see the rise of geology as a painless movement. In particular, Mary's geological friends, William Buckland and his Cambridge opposite number the Revd Professor Adam Sedgwick, were personally criticised for their views in *The Times* in 1845. We are made especially aware of this because Mary Somerville, then living in Rome, wrote about the matter to her son, Woronzow Greig, she having read of the attack in the English newspapers. Even so, she dismissed these attacks on the geologists because 'their adversaries write such nonsense and it matters little'.[46]

The anti-geological backlash—no doubt triggered by the publication of *Vestiges* —had been set in motion at York in September 1844, when the British Association for the Advancement of Science held its annual meeting in that city. The Dean of York Minster, the Revd Dr. William Cockburn, had openly attacked the geologists in an embarrassing address in which he defended strict Biblical literalism as against the new geological time scales; and some years later, following the publication of

[45] Robert Chambers, *Vestiges of the Natural History of Creation* (London, 1844, Leicester University Press reprint, 1969). See pp. 165–190 for Chambers' idea of the chemical and electrical origins of the first living things.

[46] Mary Somerville to Woronzow Greig (her son), Rome, 3 August 1845, reprinted in Mary Somerville, *Personal Recollections* [n. 5], 275–276.

her *Physical Geography* (1848), Mary herself 'was preached against by name in York Cathedral'.[47]

Mary Somerville was never a practising geologist or a geological writer, but it is clear that if she was being personally denounced from the pulpit of York Minster, then her interest in and sympathy with the new science must have been very well known. Yet within that still small world of Victorian science, her fame was already great, as a female mathematician who admired the work of Buckland, Sedgwick, Murchison, Lyell, and other geologists; and as her private remarks about the science were no doubt well known, one can fully understand how her radical views could win censure. Even though she was living in Rome in the mid-1840s, her subsequent approval of *Vestiges*, communicated by letter to her son Woronzow Greig and friends at home,[48] can only have confirmed her *avant garde* views within the intellectual community.

The aspects of geology which seemed to have particularly appealed to Mary, however, were the geophysical rather than the palaeontological. This may have been occasioned in part by the fact that she was by then a middle-aged (albeit very fit) lady, caring for an older husband who was no longer well, and moving between various Italian cities. Quite simply, she lacked the time and was not in a position to partake in the painstaking location and analysis of rock strata that was fundamental to fossil geology—not to mention the danger occasioned by the gangs of bandits who roamed parts of Italy, and to whose menacing presence she sometimes referred.[49]

I would suggest, however, that Mary Somerville's scientific instincts were predominantly those of the exact physical rather than the comparative natural history scientist. What really interested her was the Earth as a physical or planetary body, rather than as a habitat for changing life-forms. As a young woman in London, for instance, she had taken private lessons in mineralogy from a Mrs.

---

[47] Mary Somerville, *Personal Recollections* [n. 5], 129. Dr. Cockburn's Address to the British Association was not published as a part of the Association's official proceedings, but under Cockburn's own imprimatur, as *The Bible Defended Against the British Association* (London, 1844). Cockburn argued that the complex strata and palaeontological evidences that were part and parcel of academic geology by 1844 did not, in fact, signify pre-*Genesis* time scales, but were the products of divine miracles taking place within relatively recent history. His position is not dissimilar to that of certain modern-day American Fundamentalist groups. A vivid account of Cockburn's reception was given in a letter from Richarda Airy, wife of the Astronomer Royal, as passed on to her by her husband George: 'Mr. Airy found everybody talking about the Dean of York's attack on the Association on the grounds of infidelity. Mr. Sedgwick it seems had been *hammering* him down most successfully: and the ladies had gone down in crowds to witness the execution', Lady Richarda Airy to Lady Margaret Herschel, (Greenwich), 6 October 1844. Letter in private possession of the Airy family, to whom I am indebted for the loan of this and many other family documents.

[48] Mary Somerville to Woronzow Greig, Rome, 28 May 1845, in *Personal Recollections* [n. 5], 278.

[49] Mary Somerville to J.F.W. Herschel, Naples, 26[?] September 1868, Royal Society MS, Herschel Papers HS16 [377] for reference to brigands.

Lowry—a Jewish lady who was the wife of an engraver[50] (and about whom one would like to know more). And then, Mary's friend Dr. Wollaston taught her how to use the goniometer—an instrument which he had invented to measure the exact angles at which the planes of different crystals meet. In mineralogy and crystallography in particular, one finds that exact mathematical component which so clearly appealed to her mind.

Likewise, it was the physical forces that lay at the heart of continental and mountain-building processes, volcanic eruptions (Vesuvius especially fascinated her, and she and Dr. Somerville had descended into the recently erupted crater around 1818) and the erosive powers of glaciers[51] which captured her imagination in matters of field geology. For here, one can see something that is related to gravitational physics and which is—at least potentially—amenable to mathematical expression. It is this experimental and mathematical approach to scientific knowledge which was to run through all of her published work.

References

1. Armitage, A. (1962). *William Herschel* (pp. 88–94). London: T. Nelson & Sons.
2. Hoskin, M. (1997). *Cambridge Illustrated History of Astronomy* (pp. 235–237). Cambridge: Cambridge University Press.
3. Chapman, A. (1989). William Herschel and the Measurement of Space. *Quarterly Journal of the Royal Astronomical Society, 30*, 399–418.
4. Grant, R. (1852). *History of Physical Astronomy* (pp. 561–562). London: Forgotten Books.
5. Brown, E. (1890, November). Programmes of the directors of the observing sections. Solar section. *Journal of the British Astronomical Association, 1*(2), 58–60.
6. Somerville, M. (1873). *Personal Recollections* (p. 219). London: Forgotten Books.
7. Somerville, M. (1836). *On the Connexion of the Physical Sciences* (p. 412). London: Forgotten Books.
8. Chapman, A. (1998). *The Victorian Amateur Astronomer. Independent Astronomical Research in Britain 1820–1920* (pp. 93–95). Chichester and New York: Praxis-Wiley.
9. Somerville, M. (1836). *On the Connexion of the Physical Sciences* (3rd ed.) (p. 357). London: Forgotten Books.
10. Somerville, M. (1836). Extrait d'une lettre de Mme. Sommerville [*sic*] à M. Arago—Expériences sur la transmission de rayons chimiques du spectre solaire, à travers différents milieux. *Comptes Rendus Hebdomaines de sciences de l'Académie des Sciences, 3*, 473–476.
11. Rudwick, M. (1972). *The Meaning of Fossils. Episodes in the History of Palaeontology*, Chapter 3. London, New York: MacDonald, American Elsevier.

---

[50] Mary Somerville, *Personal Recollections* [n. 5], 106.

[51] Mary Somerville, *Personal Recollections* [n. 5], 125.

Mary Somerville to J.F.W. Herschel, Naples, 12 November 1868, Royal Society MS, Herschel Papers HS16 [375] refers to Vesuvius, which had been in eruption for nearly 4 months. In *Physical Geography*, 2nd edn. (London, 1849), Chapters 1 and 2, there is extensive discussion about volcanic and related forces.

Chapter 4
Mary Somerville: The Writer

Mary Somerville already had an established international reputation as a physical scientific thinker long before she actually published her first words in 1826. As we have seen, it was through the verbal and epistolary channels of the age that she had first won fame in Edinburgh, London, Paris, Geneva, and beyond. And while she was by no means the only woman to publish on science in that age, she was unique in her approach. Caroline Herschel, for instance, while a distinguished cometographer, wrote very little and did her best to avoid being exposed in society. And then there were Maria Edgeworth and Jane Marcet, whose works were aimed at the more elementary communication of science to children and young people.[1] Mary Somerville, on the other hand, wrestled with the most complex parts of the most abstruse science of her day—celestial mechanics—to produce reworkings and refinements of the ideas of figures such as Laplace. It was this phenomenal intellectual capacity and confidence, combined with a likeable modesty and charm of personal demeanour, which first established her European celebrity.

Living in an age given over to romantic excess and emotional superfluity, moreover, she displayed that hard-headed, no-nonsense approach to life which was later to be commented upon by Ellen Mary Clerke, her original *Dictionary of National Biography* biographer. Indeed, one might suggest that we find here those components of intellect and personality which so fascinated her contemporaries. To the people of Georgian England, Mary must have seemed a woman of paradoxes— a female intellectual who was decidedly *not* a 'blue stocking'; a beauty and a charmer who was also a creative higher mathematician; and a devoted wife and

---

[1] Mary Somerville mentions her friendships with Maria Edgeworth and Jane Marcet in *Personal Recollections* (London, 1873), 114 and 156. Maria Edgeworth, moreover, left a graphic description of Sir John Herschel operating his 20-foot reflecting telescope in the dark, during which 'Herschel runs up and down the ladder like a cat (because I would not say a monkey)', in her 1831 account to Harriet Butler, reprinted in Christina Colvin, ed., *Maria Edgeworth. Letters from England 1813–1844* (Clarendon Press, Oxford, 1971), 506. See also Mary T. Brück, 'Maria Edgeworth: Scientific 'Literary Lady'', *Irish Astronomical Journal*, 23, 1 (1996), 49–54.

© The Author(s) 2015
A. Chapman, *Mary Somerville and the World of Science*,
SpringerBriefs in History of Science and Technology,
DOI 10.1007/978-3-319-09399-4_4

mother to whom her distinguished physician husband was proud to act as an amanuensis.

It was, no doubt, this established reputation which secured the publication of her first paper—on the magnetising properties of the violet and 'chemical rays' of the solar spectrum (discussed in Chap. 3)—in the prestigious *Philosophical Transactions of the Royal Society* in 1826. However, a careful reading of this and her subsequent research papers (*Comptes Rendus* 1836, and *Philosophical Transactions of the Royal Society* 1846)[2] provides several clues about Mary Somerville as both a scientist and as a writer. Perhaps the most obvious of these is her remarkably mature grasp of how the scientific method operates, and how a carefully controlled set of experimental procedures could be used to uncover the inner workings of nature. Another intellectual attribute which emerges most clearly from her writings, moreover, is her controlled pragmatism and eschewal of speculation regarding the matter of what these 'chemical rays' might be. When she found it impossible to extract any clear conclusions from her optical experiments, she was content to simply put them on record in the hope that they would be of value to some future researcher. There was no tendency to speculate or draw conclusions that strayed one inch beyond the experimental results.

One of the most important traits to emerge from her three research papers, as also from her surviving letters, however, is mental clarity and precision of thought. This gift would prove invaluable when it came to writing her four books, which between them display not only a great and focused intellectual energy and power of analysis, but also a capacity for creative synthesis. Whether she is writing upon gravitational mechanics, the Earth's continents, the microscopic realm, or the 'connexion of the physical sciences', it is evident that her mind is of truly remarkable range and erudition. Running through both her published works and her correspondence is the perception of science as a body of rational, public knowledge that operates in an intellectual market place that should be independent of bias, privilege or dogma. Furthermore, this science was believed not only to reveal beautiful truths about nature, but to operate in tandem with these progressive emancipatory movements which would take the human race from darkness to light. And while we, 170 years later, might consider her vision of science to be somewhat roseate in its purity, it was completely in harmony with the noblest sentiments of the age in which she lived.

When Lord Henry Brougham besought Dr. Somerville to encourage his wife to write an expository treatise on the works of Baron Laplace in 1827, he may not have realised what forces his request would unleash[3]—or probably he *did* realise it. Even so, when *On the Mechanism of the Heavens* saw the light of day in 1831, it caused a sensation, and Mary Somerville became a celebrity scientist.

By any standards, *Mechanism* is a complex work which demands a knowledge of quite advanced mathematics for it to be truly comprehended; and considering

---

[2] See present work, Chap. 3, notes 27 and 31.

[3] Henry, Lord Brougham and Vaux to Dr. William Somerville, 27 March 1827, reprinted in Mary Somerville, *Personal Recollections* [n. 1], 161–162.

Mary's own remarks about the generally assumed intellectual capacities of women in her day, one can understand why it created such a sensation.

Dedicated to her friend and encourager, 'Henry, Lord Brougham and Vaux', it pulls no punches, and makes it clear from the outset in that celebrated 'Preliminary Dissertation' which preceded the mathematical body of the text, and which became a major statement on the intellectual power of science in its own right, that 'A complete acquaintance with Physical Astronomy can only be attained by those who are well versed in the higher branches of mathematical science: such alone can appreciate the extreme beauty of the results, and of the means by which these results are obtained.' So, one might say, the reader is warned from the start that this is not going to be a cosy read, and that without mathematics an understanding of the inner workings of astronomy is impossible. Her friend Joanna Baillie further commented that her book and 'Preliminary Dissertation' had done much 'to remove the light estimation in which the capacity of women is too often held'.[4]

Central to the *Mechanism* (as would be the case with the *Connexion* in 1834) is Mary Somerville's concept of the intellectual unity of physical science. What is more, this unity hinged upon the classical truths of mathematics, and in particular the seemingly all-embracing power of gravitation theory; for if one could express the behaviour of matter and motion through the equations of the inverse square law of Newtonian gravitation, then one could create a truly comprehensive and explanatory model of science. Indeed, we can perhaps see here why Mary appeared to show less interest in the more comparative or taxonomic sciences of botany or palaeontology than she showed in physics: the comparative sciences, being non-mathematical, depended too much on the opinions or experiences of the individual scientists, and seemed less exact in their methods. In short, these sciences had yet to find their Newton, who would reduce them to exact laws.

In the *Mechanism*, Mary goes on to explore gravitation theory both conceptually and in its application to observable phenomena. As a preliminary to this process, the establishment of techniques and standards of exact measurement was fundamental. In consequence, she next discusses early chronological systems and the recent archaeological discovery of an ancient Egyptian zodiacal division of the sky, as well as the French Revolutionary metric division of time and other physical units.

Mary then embarks upon a survey of gravitation theory, the range and thoroughness of which must have left many of her readers gasping. There is a clear discussion of Newton's concept of the gravitational point mass centre of an astronomical body, whereby in terms of orbital dynamics a planet many thousands of miles in diameter still behaves as though all of its mass were concentrated in a single central point. And at the Earth's own point mass centre at the centre of the Earth, according to Dr. Thomas Young, our planet's gravitational pull would be so

---

[4] Mary Somerville, *On the Mechanism of the Heavens* (London, 1831), Dedication; and see the 'Preliminary Dissertation', vii, which prefaces the work (and which was to win for Mary a quite separate distinction as a statement of the intellectual power of science). See also Miss Joanna Baillie to Mary Somerville, 1 February 1832, in Mary Somerville, *Personal Recollections* [n. 1], 206.

great that 'steel would be compressed to one-fourth and stone into one-eighth of its bulk'.[5]

This in turn leads to an analysis of what happens when gravitational forces come into contact with each other, such as those between the Earth, Sun, Moon and other Solar System bodies. These combined forces produce the strangely-shaped cometary orbits, the Earth's tides, and even the isochronal beats of a clock pendulum, as the period of swing of the pendulum is an exact function of both the pendulum's length and the Earth's gravitational pull for a given location on the Earth's surface.[6]

Her book book then really gets down to business, as she starts to define the axioms that underlie mathematical astronomy and the effects of gravity upon the orbital behaviour of solid and fluid (oceanic) bodies. She next commences her systematic study of why planetary orbits are elliptical, and how satellites such as our own Moon and those of Jupiter behave with relation to their parent body and to the Sun. Comets are examined, along with the perturbations or gravitational disturbances which planets exert both upon comets and over each other, for as planetary positions are in a constant state of change, so likewise are their mutual gravitational relationships. And all of this motion and what Newton called 'fluxion' is capable of precise mathematical expression.

It is clear, however, that once the bit was firmly between Mary's teeth and the composition of the *Mechanism*—instead of simply a condensed English version of Laplace's *Mécanique Céleste*—was under way, a quite independent creative vision came into being. For as she later made clear in her Introduction, the 'object of this work is rather to give the spirit of La Place's method than to pursue a regular system of demonstration'. In short, by working within 'the spirit of La Place's method', she was going on to produce creative science in her own right (Fig. 4.1).

The depth of mathematical exploration which Mary Somerville plumbed in her composition of the *Mechanism* is perhaps nowhere more evident than in the correspondence which she exchanged with Sir John Herschel over this time. By the late 1820s Mary regarded Herschel as her mathematical benchmark of authority, and it was due to his enthusiastic support that she was encouraged to explore these concepts and algebraic expressions. Her regard for Herschel stemmed not just from his encouragement, however, for she also respected the power of his objective criticism, knowing as she did that if the sheets of text and equations which she sent to him earned his approval, then they would also pass any scrutiny in Europe. Indeed, in a letter to Mary of 23 February 1830, Herschel styled himself 'a rough critic—but I think of Horace's *good critic*', though as she responded, on 15 May, she knew that her ensuing work would not go to print 'contain[ing] any great blunders' if Herschel had read it.[7]

---

[5] Mary Somerville, On the Mechanism of the Heavens [n. 4], Preliminary Dissertation, xl.

[6] 6. Mary Somerville, On the. Mechanism of the Heavens [n. 4], Preliminary Dissertation, xxvii. Also Physical Astronomy: 'Introduction' (which follows 'Preliminary Dissertation', pp. 1–3).

[7] John F.W. Herschel to Mary Somerville, 23 February 1830, Royal Society MS, Herschel Papers HS16 [329]. Also, Mary Somerville to Herschel, 15 May 1830, Royal Society MS, Herschel Papers HS16 [336].

Fig. 4.1 Sir John Frederick William Herschel (1792–1871), only child of Sir William and Lady Mary Herschel, and Mary Somerville's great encourager. (Royal Astronomical Society, Presidential Portrait 4C.)

On the Mechanism of the Heavens was an instant success, and became the foundation stone on which Mary's subsequent reputation would stand. Its original print run of 750 copies sold quickly, and Jean Baptiste Biot reviewed it very favourably for the Académie des Sciences in Paris; but perhaps the greatest accolade came from Cambridge, when in November 1831 William Whewell, the future Master of Trinity College, wrote, as was customary, to Dr. Somerville, praising the book. Whewell then went on to use *Mechanism* as a teaching text for his mathematical Tripos students, thereby making it the first scientific textbook written by a

---

[8] Dr. William Whewell to Dr. W. Somerville, 2 November 1831, reprinted in Mary Somerville, *Personal Recollections* [n. 1], 170–171. Also Professor George Peacock to Mary Somerville, 14 February 1832: *Personal Recollections*, 172.

woman to be used in a British university. Mary considered this to be 'the highest honour I ever received'.[8]

On the other hand, it must not be forgotten that the thought of a woman presuming to amplify the work of Laplace aroused derision in some quarters. Charles Buller, a West Country MP, openly mocked Mary's efforts before the House of Commons, and she was rebuked for her presumption and ignorance by an unnamed individual at a party at Lansdown House; while it was most likely Buller again who wrote the unsigned scathing review of *Mechanism* in the 21 January 1832 number of the *Athenaeum* magazine. The *Athenaeum* reviewer, indeed, not only poured scorn upon a woman grappling with Laplace, but also upon the wider intentions of Lord Brougham's Society for the Diffusion of Useful Knowledge, to take science to a wider and less elite readership.[9]

But if *Mechanism*, with its elegant English phrases and minutely argued pages of mathematical equations, showed off Mary Somerville's brilliance as a gravitational physicist, so her next book, *On the Connexion of the Physical Sciences* (1834), displayed her command of a wide range of scientific knowledge. The purpose of *Connexion*, in fact, was to guide the reader through the various branches of contemporary physical science, and to demonstrate how one integrated set of intellectual principles could be traced through all of them. It did so, however, without equations, which made it accessible to a very much larger readership than the undergraduates and already mathematically literate lay readers of *Mechanism* (Fig. 4.2).

Connexion is ultimately about the interrelationship of quantifiable physical forces. There is, of course, much discussion of gravitation, in which Mary returns to the same broad topics which she had dealt with in *Mechanism*, but now expounded purely in words. Orbital shapes, the physical cause of the Earth's oblateness, the tides, measurement of the dimensions of the Solar System, binary stars and Herschel's cosmology are all treated with great detail. And as in *Mechanism*, she pays tribute not only to the pioneering researches of Sir William Herschel, but also to his sister, 'Miss Caroline Herschel, a lady so eminent for astronomical knowledge and discovery'.[10]

In addition to dealing with the broader gravitational environment, however, *Connexion* addresses those still imperfectly understood terrestrial forces of light, sound transmission, static and current electricity, Faraday's electromagnetic

---

[9] *Athenaeum* 221 (1832), 43–4. The critical attacks on *Mechanism* are discussed by Mary Patterson, *Mary Somerville and the Cultivation of Science 1815–1840* (Martinus Nijhoff, Kluwer Group, Boston, The Hague, Lancaster, 1983), 84–85. Mary Somerville's own responses to Charles Buller's and other slights were noted in her autobiographical draft but subsequently omitted from the published *Personal Recollections*. They are included in the new scholarly edition of her text, *Queen of Science. Personal Recollections of Mary Somerville*, ed. Dorothy McMillan (Canongate Classics 102, Edinburgh, 2001), 145–146.

[10] Mary Somerville, *On the Connexion of the Physical Sciences*, 3rd edn. (London, 1836), 397. For the Caroline Herschel reference in *On the Mechanism of the Heavens* [n. 4], see 'Preliminary Dissertation', lxvi.

Fig. 4.2 (*Upper*) Jupiter, with its complex system of belts and zones, in 1832; (*lower*) Saturn and its rings, showing the inner crêpe (C) ring discovered in 1850 (J.F.W. Herschel, *Outlines of Astronomy*.)

researches, and magnetism. Light, in particular, was one of Mary Somerville's great interests (as it was also one of Sir John Herschel's), and she makes no bones of her partisanship of Dr. Thomas Young's 'undulatory' (wave) theory of light as against the older Newtonian 'corpuscular' (particle) theory (as discussed in Chap. 3). She then examines Herschel's 'calorific' (infrared) rays and the 'chemical' (ultraviolet) rays of the solar spectrum discovered by von Ritter and others.

Although no-one could be certain of the fact in the 1830s (although many suspected it), light seemed closely connected with electricity, and in Sect. XXVIII of *Connexion*, Mary examines the contemporary state of knowledge of electricity seen in its context as a force of nature. She discusses contemporary theories (primarily Faraday's) about the generation of 'Voltaic' (current) electricity in chemical batteries, and its possible relation to the oxidation of metals by acids, before considering the contemporary idea that torpedo, ray and other electrical fish generate their shocking power from anatomical features in which certain structures seem to be separated by fluid in their bodies—in a way that is rather similar, in fact, to the acids and metals in a chemical battery. Electricity is next discussed with reference to the Aurora Borealis (the Northern Lights), and she cites experimental evidence from Faraday and Arago concerning Earth currents.

The latter part of *Connexion* returns to celestial phenomena, as Mary draws upon evidence in contemporary scientific literature to argue in favour of a tenuous physical 'ether' pervading the whole of space. Surely, in the same way that the Earth's atmosphere conveys sound by means of waves, so such a cosmological ether must exist to convey the undulating waves of light. And was this ether in some way related to those seemingly wispy nebulae discovered in deep space by the Herschels? It is true that when she comes to discuss the very constitution of space and the bodies that occupy it, one encounters what looks like guarded speculation. Yet she is no more speculative in these respects than were the Herschels, Faraday, Arago, or any of the other research scientists of the age who tried to explain classes of phenomena which exhausted the vocabulary of empirical science as it was then understood.

In her Conclusion, Mary draws all of these diverse aspects of science together into a singularly elegant formulation, for 'even the imponderable matter of electric, galvanic, or magnetic fluid—all are obedient to permanent laws, though we may not be able in every case to resolve their phenomena into general principles.' The whole body of scientific knowledge which the human mind is capable of comprehending, moreover, 'rests on a few fundamental axioms, which have existed eternally in Him who implanted them in the breast of man when he created him after His own image'.[11]

On the Connexion of the Physical Sciences—combining as it did an intellectual vision of science that was truly encyclopaedic in scope, with an ease and clarity of expression—became an immensely influential book. It passed through nine editions before Mary's death in 1872—each one being carefully revised to incorporate the latest discoveries and opinions—and a tenth edition was published in 1877. And although the visionary structure of the book remained the same, its basic argument was continually being strengthened as former sections were amplified in the light of new data. The five editions published before 1840, for instance, say nothing about photography, but after the processes of Daguerre, Fox-Talbot and John Herschel were perfected around that date, photography is treated with increasing detail not only as a technique whereby the invisible rays of the solar spectrum can be further investigated, but also as an image-recording medium.[12] Similarly, her treatment of electricity expands as new evidence of electrical conductivity, as manifested in such

---

[11] Mary Somerville, *On the Connexion of the Physical Sciences* [n. 10], 412.

[12] In the ninth edition of *On the Connexion of the Physical Sciences* (1858), Section XXIV, the Daguerrotype, calotype, chromatype and other photographic processes, along with the 'chemical spectrum', are discussed at length.

practical application as telegraphic technology, became a normal part of life by the mid-Victorian age.

Connexion also inspired one of the most significant and fruitful investigations ever to have been undertaken in the whole history of celestial mechanics—an investigation which was in itself a spectacular vindication of the central argument concerning the unity and universality of scientific law which ran through both *Mechanism* and *Connexion*. This was the discovery of the planet Neptune in 1846, due to painstaking analysis of the gravitational perturbations of the already known planet Uranus. When the English mathematician John Couch Adams (who obtained his figure for the place of the then unknown Neptune several months before Urbain Le Verrier in Paris), together with the Somervilles, partook in a Christmas house-party at Sir John Herschel's mansion, Collingwood, at Hawkhurst, Kent, in 1848, the now famous Adams conveyed a remarkable confidence to Dr. Somerville. 'Mr Adams told Somerville that the following sentence in the sixth edition of the 'Connexion of the Physical Sciences' published in the year 1842 put it into his head to calculate the orbit of Uranus.' In this sentence, Mary suggested using the observed tabular errors of the motions of Uranus as a way of pinpointing the position of the unknown planet whose gravitational influence was dragging Uranus out of its orbit.[13]

In many ways it is also remarkable that Mary Somerville—who had never travelled beyond Europe, and who at the same time possessed such an experiment-based concept of science—should have tackled such a subject as *Physical Geography* (1848). And though she had felt reluctant to proceed with the book following the publication of Alexander von Humboldt's *Kosmos* (1845), her *Physical Geography* is nonetheless a great work of synthesis and vision; for here we see the familiar intellectual concepts so elegantly established in her two previous books—the unity of all physical knowledge and its susceptibility to law-like comprehension—now applied to geographical and geophysical problems.

As one might expect, her predominant interests lay with terrestrial structures rather than with living fauna, and the earlier part of the book examines mountain-building processes, as they were then understood, as a way of explaining the shapes and masses of the continents and their alignment with the Earth's rotational and orbital characteristics. She then describes the moulding of these continents by glacial and other erosive forces. True to form, however, she displays no partisanship to either Catastrophist or Uniformitarian theories of continent formation, drawing instead upon a wide and eclectic range of sources, and acknowledging Lyell,

---

[13] I have not been able to locate a copy of the sixth edition of *On the Connexion of the Physical Sciences* (1842), but the passage in which Mary Somerville referred to the possibility of using the orbital disturbances of *known* planets to discover the positions of unknown planets (cited in *Personal Recollections* [n. 1], 290), and which had first inspired Adams, appeared in previous editions: see *On the Connexion of the Physical Sciences*, 5th edn. (1840), 382. By the time of *Connexion*, 9th edn. (1858), the discovery of Neptune was discussed in some detail: 22, 62.

[14] Mary Somerville, *Physical Geography* 1, 2nd edn. (London, 1849), 44, footnote (for a list of names: Lyell, Murchison, etc.).

Cuvier, Murchison, De la Beche, Owen, and others.[14] And the Biblical Flood of Noah is conspicuous by its absence!

In addition to the globe itself, *Physical Geography* treats extensively of lakes, oceans, and the atmosphere, as the relationship between temperature, evaporation, dew, and frost are explored, and related to climatic zones. And in her thoroughgoingly instrumental approach to science, she cites numerous observations and recordings made by individual scientists and explorers on mountain tops, in oceans, and in a diversity of locations, wherever they are available.

The 1830s and 1840s were a time when scientists such as Justus von Liebig in Germany and Charles Daubeny in Oxford[15] were coming for the first time to understand the complex organic chemistry that lay behind plant growth, and in particular the roles played by gases such as carbon, hydrogen and nitrogen in that growth. In *Physical Geography*, Mary Somerville examines plant chemistry, temperature and local conditions as necessary physical prerequisites for the proliferation of fauna across the planet. Indeed, in this respect she further develops the subjects touched upon more than a decade earlier in *Connexion*, in which she dealt with the conditions necessary for the growth of exotic jungle flora, and explained why the most dangerous types of snake were found in the tropics.[16]

All of this new knowledge of plant chemistry, of course, came to be conjoined in 1849 with contemporary advances in Mary's long standing research interest in the 'invisible rays' of the Sun and their effect upon vegetable growth. In Chap. XXIII of *Physical Geography* the reader is treated to a state-of-the-art account of studies of the solar spectrum, further amplified and quantified by their measured effects on Daguerrotype photographic plates, which were believed to provide a standard of comparison that was independent of the variable perceptions of the human eye. We are told, for instance, that 'the blue rays of the solar spectrum have most effect on the germination of seed; [while] the yellow rays, which are the most luminous, on the growing plant.' In spring, the 'invisible' (ultraviolet) rays were the most abundant and most conducive to plant growth and the production of the green parts of plants, whereas the high summer Sun, shining down through less atmosphere, produced a greater abundance of red 'heat' or 'calorific' (infrared) rays, which were essential to the ripening of fruit. This photochemical discovery, indeed, had already led to changes in the design of the glass houses at Kew Gardens. The 'calorific' rays being admitted by the clear glass in the Palm House had been found to be destroying the colour of the fronds, which led to the clear panes being replaced by new ones tinted pale yellow-green with copper oxide, which filtered out the most

---

[15] F.L. Holmes, 'Justus von Liebig', *Dictionary of Scientific Biography* (New York, 1981), 329–350; 'Charles Daubeny', *Dictionary of Scientific Biography* 585–586. Nigel I. Miller, 'Chemistry for Gentlemen: Charles Daubeny and the Role of Chemical Education at Oxford', Oxford University Chemistry Part II Thesis, 1986, deposited in the Oxford History Faculty Library: 1–13, 29–44, 66–80.

[16] Mary Somerville, *Physical Geography* vol. II [n. 13], Chapters XXIII ff. See, for example, Chapter XXX, 253, for her discussion on the greater prevalence of venomous snakes in the tropics.

[17] Mary Somerville, *Physical Geography* II [n. 14], 101–102.

damaging rays.[17] As a consequence of these factors of latitude, soil and atmospheric chemistry, combined with the differing chemical influences of red and blue light, one could explain, in physical terms, why particular plants grew best in different locations.

In *Physical Geography* Mary Somerville once again displays her extraordinary capacity to absorb and master a wide range of different types of knowledge and data, discern the common physical threads running through them, and use the resulting information to build up a powerful argument. And while one might say that such a book is essentially a *secondary* source insofar as it is based on an eclectic understanding of the primary researches of others, it is still an extraordinary piece of scientific interpretation that could have been produced only by a person possessing the synthesising insight to see the global picture. And like her previous books, it was very favourably received by both the reviewers and the public. It had passed through six editions by the time of her death in 1872, and was to run into a seventh in 1877.

After *Physical Geography*, published when she was 69 years old, Mary Somerville was to produce no new work for 20 years. Dr. William Somerville died in 1860, and then, most tragically, Mary's only surviving son, Woronzow—who was then practising as a barrister in London—died in 1865. Of course, she revised the earlier editions of her already famous three books, corresponded from her home in Italy with friends and scientists across Europe, and kept herself thoroughly up to date in her reading on the progress of science. Yet what is truly remarkable is that a lady of such a great age should still possess the vision and the energy to produce another major book that would finally come off the presses when she was 89. *Microscopic and Molecular Science*, published in two substantial volumes in 1869, deals with the structure of matter, and in particular, that of living organisms. It draws upon and gives acknowledgement to a range of eminent scientists—many of whom, moreover, would have been only boys when she first published *On the Mechanism of the Heavens* 48 years earlier: men such as John Tyndall, James Prescott Joule, Gustav Kirchhoff, Sir William Huggins and Warren de la Rue; while William Perkin—pioneer of the organic chemistry of coal-tar dye-stuffs—had not even been born.

Volume I of *Microscopic and Molecular Science* addresses itself to theories of the nature of atoms and molecules. However, although the book incorporates a great amount of recent research, its overall approach makes much of its science seem somewhat old-fashioned, and John Murray the younger—whose father's publishing house had issued all of Mary's books—published it out of respect and loyalty as much as for scientific relevance.[18] Even so, Mary had carried out her preparation thoroughly, although by this date the sciences were advancing so rapidly that even the ever-resourceful Herschel expressed amazement at the rate of progress. Modern theories of atoms and their suggested potential in creating living structures quite made Herschel gasp, and as he said in one of his letters to Mary: 'The idea seems to

---

[18] See Ref. [1].

be going around that the Universe is a mere collection of Billiard Balls knocking one another by *simple collision*!!', along with 'all the phaenomena of reproductive life and omni animalium' (these last words are not clear in the letter). Indeed, 'It's enough to puzzle a Senior Wrangler of 1865!'[19]

Theories about how molecules were believed to hold together are then reviewed in Mary's text, along with contemporary experiment-based ideas about the inner forces of nature, such as heat and electromagnetism. Joule's work on thermodynamics and energy conservation are examined, along with the rise of organic chemistry, and what they can teach us about the inner constitution of matter. One of the most dramatic revelations of organic chemistry was that living organism-derived substances were not, in their building blocks, chemically and physically different from inorganic substances. The chemical building blocks or elements—carbon, hydrogen, oxygen, nitrogen, and so on—could be the same, although the molecular bonds between these elements were much more complex than was usually the case with inorganic substances. What chemists had shown, indeed, was that organic compounds such as blood, coal tar and milk did not possess some unique 'life force' which made them special. They simply had a more complex molecular structure.

Attention is also paid to a rapidly maturing science which had made its first infant steps when Mary was young, and in which she had published her own research papers of 1826, 1836 and 1846: spectroscopy. In addition to the earlier reported work on the Sun's 'chemical' and 'calorific' rays, scientists such as Sir David Brewster, Léon Foucault, Gustav Kirchhoff, Count Robert Bunsen and William Huggins had, after 1850, made a series of discoveries and advances which revealed the spectroscopic analysis of light to be a wonderfully sensitive technique whereby complex chemical substances could be analysed with the greatest accuracy; and after 1859, Kirchhoff, Bunsen and Roscoe in Heidelberg had shown that not only could the spectroscope be used as an analytical technique in the laboratory, it could even detect and quantify the chemicals present in the Sun and stars.[20] Indeed, in the same vein as Kepler and Newton, Kirchhoff went on to devise his three Laws of Spectroscopy.

It is clear that Mary Somerville was fascinated by the power of the spectroscope, as in many ways it afforded a perfect vindication of her whole life's work as a scientist. Just as *Mechanism* had explored the universality of gravitation, from pendulums to the Solar System, and *Connexion* had showed that light was believed (but not yet proved) to be an all-pervading aspect of electromagnetism, so the spectroscope had shown that the same fundamental substances which made up human bodies were also to be found in incandescent astronomical objects. The

[19] John EW. Herschel to Mary Somerville, 11 April 1865, Royal Society MS, Herschel Papers HS16 [372].

[20] Owen Gingerich, 'Unlocking the Chemical Secrets of the Cosmos', in Gingerich, *The Great Copernicus Chase* (Sky Publishing, Cambridge, Massachusetts; and Cambridge University Press, 1992), 170–176.

spectroscope, indeed, demonstrated that one great, universal, and law-like set of chemical and physical structures ran through the entire cosmos.

In Volume II of *Microscopic and Molecular Science*, Mary's writing still follows an essentially mathematical and chemical rather than a comparative botanical approach. What is more, it is evident that the things which most appealed to her in the realm of living beings were their chemical and physical structures, for 'Carbon, and the three elementary gases [presumably oxygen, nitrogen and hydrogen] constitute the bases of all.'[21] She is especially concerned with how basic chemical building blocks, when combined with heat, produce proteins and other complex organic substances. Likewise, in the microscopic realm, her particular interest lay in how those substances, and the cells which were built up from them, were always arranged in patterns of exquisite mathematical beauty.

In 1835, Europe and America were set agog by the predicted return of Halley's comet. The excitement derived in part from the great sweeping beauty of the comet's tail, yet as this was only the second time that the comet had returned since Edmond Halley had calculated that the comet of 1682 was periodic (it had returned in 1758), its obedient reappearance in 1835 seemed only to confirm the wonder of those exact laws of gravitation which controlled its orbit. And no doubt because of her increasing celebrity as the author of *Mechanism* and *Connexion*, Mary Somerville was asked by the publisher John Murray to write an essay article upon it (unsigned, by the conventions of the day) for the influential *Quarterly Review*.[22]

In this article, we see Mary Somerville in her new role not as a research scientist, but as a scientific essayist writing for the quality-magazine-reading public. *Halley's Comet* was an elegant piece of scientific writing, painlessly free of mathematics, yet explaining to a lay readership what comets were known to be by 1835, and the laws which they obeyed. Her article gave rise to a private correspondence with her fellow Jedburghian amateur astronomer, James Veitch, regarding the discovery of comets by amateur as opposed to professional astronomers. In *Halley's Comet*, Mary had mentioned the historical point that while all the astronomers of Europe were watching the skies towards the end of 1758, to see if the comet predicted by Halley would return, the honour of discovery had actually fallen to an Hungarian peasant astronomer. But as Veitch had reminded Mary, she had failed to mention that the honour of discovering 'the comet of 1811, the greatest that had appeared for a

---

[21] Mary Somerville, *Microscopic and Molecular Science*, vol. I (London, 1869), 167.

[22] *Quarterly Review* CIX, VII (December 1835), 195–233. Rather strangely, Mary Somerville's name is nowhere associated with the article on Halley's comet, which is listed as Section VII, in the *Quarterly Review* (December 1835), 195–333. Indeed, the articles printed in those pages are headed '1. Ueber den Halleyshen Cometen. Von Littrow. 2. Ueber den Halleyshen Cometen. Von Professor von Encke'. Nonetheless, in *Personal Recollections* [n. 1], 100, Mary Somerville mentions having been invited by John Murray to write an article on Halley's comet for the *Quarterly Review*, while James Veitch, in his letter to Mary, seems to be referring to an article written by her which also appeared in the December 1835 *Quarterly Review*: James Veitch to Mary Somerville, 12 October 1836, *Personal Recollections* 101.

[23] James Veitch to Mary Somerville, 12 October 1836, in Mary Somerville, *Personal Recollections* [n. 1], 101.

century', had fallen to 'a peasant of Inchbonny',[23] Jedburgh; for Veitch himself, who claimed its first sighting, was a Scottish amateur astronomer who earned his living making and using agricultural implements. Mary therefore set straight the public record, and in her *Personal Recollections*[24] she gave full acknowledgement to Veitch, whom she clearly knew very well.

But beyond the invitation of Lord Brougham and her friendships with many scientists of eminence, what was it that motivated her to write four major books of scientific interpretation? That the books must have been financially remunerative is beyond doubt, especially as the second and third passed through many editions during her lifetime. *Mechanism* came out in only one edition, but *Connexion* went through nine editions and sold 13,500 copies, and six editions of *Physical Geography* were published during her lifetime; and thereafter they went through a tenth and a seventh edition respectively in 1877. Nor, of course, did this include foreign (mainly American) 'pirated' editions, which certainly added to her reputation, but which in the absence of international copyright laws failed to earn her a penny. Yet in her *Personal Recollections*, Mary makes clear the source of her motivation: 'I must say that profit was never an object with me: I wrote because it was impossible for me to be idle.'[25]

Indeed, it is this driving energy, and a constant need to explore and to communicate, that runs through all of Mary Somerville's publications and achievements. And when considering this relentless curiosity combined with great physical stamina and a practical approach to life, there emerges an understanding of what it was that carried her through 92 busy years.

Reference

1. Patterson, E. (1979). *Mary Somerville 1780–1872* (p. 38). Oxford: Bocard & Church Army Press.

---

[24] Veitch, alas, had not been the first astronomer to sight the comet of 1811, as that honour went to a M. Flaugergues at Viviers in the Rhone valley in France, who saw it on 26 March 1811. Veitch had corresponded about it with Sir David Brewster, 5 August 1812: reprinted in *The Home Life of Sir David Brewster, by his daughter Mrs Gordon* (Margaret Maria) (Edinburgh, 1869), 78–79. For the discovery of the comet of 1811, see also J.R. Hind, *The Comets. A Descriptive Treatise Upon these Bodies* (London, 1852), 110. For Veitch, see A. Chapman, *The Victorian Amateur Astronomer. Independent Astronomical Research in Britain 1820–1920* (Praxis-Wiley, Chichester and New York, 1998), 185–188.
[25] Mary Somerville, *Personal Recollections* [n. 1], 202.

Chapter 5
A Full, Rich Life

In 1838 Dr. William Somerville suffered an almost fatal attack of jaundice (a disease of the liver), though in 1812, while on honeymoon in the Lake District, he had been struck down with a dangerous fever, from which it took him a month to recover. Today we may live in fear of physiologically degenerative diseases such as cancer, but rarely feel seriously threatened by a person-to-person infection, and it is all too easy for us to forget the menace of infectious disease which was faced in the early nineteenth century. Even those who moved in the highest echelons of intellectual society, who were pioneering the thought of the age, still lived in dread of scarlet fever, typhoid, typhus and other diseases. The sudden death of Mary Somerville's first husband, Captain Samuel Greig, at the age of 29, was almost certainly due to infection, while the deaths of her own small children and of many of her friends recorded in her letters were likewise due to infectious diseases. And as the cause and vector of transmission of all of these diseases were quite misunderstood before the 1860s, even distinguished physicians such as Dr. Somerville were just as vulnerable to them—and as helpless in effectively treating them—as was everyone else.

As a result of his illness, his slow convalescence, and possible damage to his otherwise strong constitution, the 69 year old William Somerville retired from his post as Physician at the Chelsea Hospital in 1840. And as he had also suffered serious financial losses as a result of a bad investment in 1835, the Somervilles decided to leave England for the milder climate and much cheaper cost of living to be found in Italy.[1] With the exception of visits to England in 1844 and 1848, when they spent time with the Herschels, Italy was to be the home of the Somervilles for the rest of their lives. William Somerville died in 1860 and was buried in the English Cemetery in Naples, and Mary followed 12 years later.[2]

---

[1] See Ref. [1].

[2] Mary Somerville, *Personal Recollections* (London 1873), *post mortem* note by Martha Somerville: 'Her remains rest on the English Campo Santo of Naples' 377.

© The Author(s) 2015
A. Chapman, *Mary Somerville and the World of Science*,
SpringerBriefs in History of Science and Technology,
DOI 10.1007/978-3-319-09399-4_5

Even so, there is no evidence that they felt at all isolated. William Somerville recovered his health in Italy, and after 17 years' residence in that country, and in his 86th year, Mary was able to inform Herschel that 'Mr. Somerville, for his age he is wonderfully well, takes a long walk every fine day, (and) goes to the Club to hear the news.'[3] The Somervilles also enjoyed a good social life, including the Italian delights of the opera and carnival; and the Italians clearly regarded Mary as a national asset, according her various scientific honours such as Honorary Membership of the Royal Italian Geographical Society, and the receipt of that Society's first Gold Medal in 1870.[4] She was also seen as an inspiration for Italian women intellectuals, and on 28 May 1853, Countess Caterina Bon-Brenzoni wrote to Mary to inform her how she was much respected by Italian women.[5] It is further clear that she, in her increasingly 'celebrity' role, was very pleased to receive English visitors —especially if they were old friends, relatives of friends, or scientists. One of the sons of Charles Babbage (the inventor of the Difference Engine) happened to visit the Somervilles in 1845, when working in Italy as an engineer overseeing the construction of a railway between Genoa and Milan.[6] This was probably Benjamin Herschel Babbage, who in 1840 had gone out as Brunel's pupil to work on the railway, though it could also have been his younger brother, Dugald Babbage, who also worked as a civil engineer on the construction of the same line.[7] And when the British Mediterranean Squadron paid a friendly call at the Italian naval base at Spezia, down the coast from Genoa, Mary visited HMS *Resistance,* which happened to be commanded by her nephew, Henry Fairfax. *Resistance,* in fact, was one of the Royal Navy's new iron-clad battleships, and the octogenarian Admiral's daughter was delighted to receive a guided tour of the vessel's engine rooms, screw propulsion system, and other new technical features (Fig. 5.1).[8]

Although Mary possessed a modest reflecting telescope, made for her by James Veitch, there is no indication in her writings or correspondence that she was a regular or systematic observational astronomer.[9] One wonders, therefore, how

---

[3] Mary Somerville to John F.W. Herschel, Florence, 14 April 1857, Royal Society MS, Herschel Papers HS16 [359].

[4] Mary Somerville, *Personal Recollections* [n.2], 301, 351. She also received the Victoria Medal of the British Royal Geographical Society, p. 350.

[5] Caterina Bon-Brenzoni to Mary Somerville, Verona, 28 May 1853: Mary Somerville, *Personal Recollections* [n.2], 297–309. See also Martha Somerville's note. A translation of Countess Bon-Brenzoni's letter is printed in *Queen of Science. Personal Recollections of Mary Somerville,* ed. Dorothy McMillan (Canongate Classics 102, Edinburgh 2001).

[6] Mary Somerville to Woronzow Greig (her son), Rome, 3 August 1845, reprinted in *Personal Recollections* [n.2], 275–276.

[7] Maboth Moseley, Irascible Genius. A Life of Charles Babbage (Hutchinson, London, [2]) pp. 167.

[8] Mary Somerville, *Personal Recollections* [n.2], 332.

[9] Mary Somerville, *Personal Recollections* [n.2], 99, for her ownership of a Veitch telescope: '… a very small one: it was the only one I ever possessed.' The subsequent whereabouts of this telescope is unknown. She also had a microscope by Adie and Sons of Edinburgh: see T.N. Clarke, A.D. Morrison-Low and A.D.C. Simpson, *Brass and Glass. Scientific Instrument Making*

Fig. 5.1 William Parsons (1800–1867), Third Earl of Rosse, builder of the great 72 inch telescope. He corresponded with Mary Somerville when she was living in Italy (R.S. Ball, *Great Astronomers.*)

envious her astronomical friends back at home must have been of the clear skies under which Italy basked. In 1844, Lord Rosse—who had built the largest reflecting telescopes of the nineteenth century, with mirrors of 36 inches and 72 inches diameter respectively, at a cost in excess of £12,000—told Mary that he was glad if he could get skies of sufficient clarity to use his great telescope to optimum advantage for a mere 30 h/year at his great observatory mansion at Birr, in central

---

(Footnote 9 continued)
Workshops in Scotland, as illustrated by instruments from the Arthur Frank Collection of the Royal Museum of Scotland (National Museum of Scotland, 1989), p. 22 Ref. [31] for Veitch telescope; p. 63 Ref. [273] for Adie microscope.

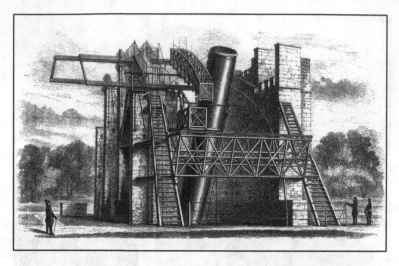

Fig. 5.2 The Leviathan of Parsonstown (now Birr), in central Ireland. This giant telescope—designed, built, and paid for by Lord Rosse—was the biggest telescope in the world when it was completed in 1845. Its mirror was 72 inches in diameter, and it focal length was 52 feet. (G.F. Chambers, *A Handbook of Descriptive and Practical Astronomy*)

Ireland.[10] Likewise, Sir John Herschel, living deep in rural Kent, often suffered poor skies and adverse conditions for several weeks in succession. And yet, as Mary told Sir John on 12 November 1843, she often found herself to be living under skies of the most wonderful unpolluted transparency. During the summer of 1842, for example, Mary had been in Venice where, so she told Herschel, 'the brightness of the sky and the clearness of the water' was such that 'you might have made your observations on the double stars and nebulae (reflected) in a nether sky in which the lustre of the Milky way (sic) seemed to be hardly diminished' (Fig. 5.2).[11]

On the other hand, she found that her gender made it impossible for her to gain access to the 8-inch aperture refractor (with an object glass made by Robert Cauchoix) in the Jesuit-operated Collegio Romano Observatory, when she wished to observe the comet of 1842. She was also requested by John Herschel to give him an account of that same instrument with which Father de Vico had done so much valuable work (Herschel scarce credited that de Vico could see as much as he claimed with an 8 inch aperture telescope), though he suspected that 'female eyes'

[10] For the cost of Lord Rosse's telescope, see Thomas Woods, *The Monster Telescope* (1845), 4. The same sum is given in John Pringle Nichol's article 'The Wonders of the Telescope', *c*.1850, a single cutting of which is preserved in Dr. John Lee's 'Scrapbook' No. 4: Museum of the History of Science, Oxford, Gunther 37, Vol. 4, 42. For Rosse's 30 h/year of good skies, see Earl of Rosse to Mary Somerville, 12 June 1844, reproduced in *Personal Recollections* [n.2], 215–216.

[11] Mary Somerville to John F.W. Herschel, Rome, 12 November 1843, Royal Society MS, Herschel Papers HS16 [347].

were prohibited access to the Jesuit observatory. Herschel then asked if Dr. Somerville had ever seen through the instrument, for 'On his report I know I could quite rely.'[12]

Through a combination of personal and intellectual qualities, and the remarkably open character of British Grand Amateur scientific society, Mary Somerville came to enjoy an illustrious reputation as a physical scientist—a reputation which spread, moreover, into the much more professionalised science of continental Europe. Yet what was Mary's attitude towards other women, and to their aspiration towards scientific and higher education? As we have already seen, she held Caroline Herschel in the highest regard, and was genuinely touched to be thought of as an inspiration to Italian intellectual women. However, when looking beyond particular cases, one senses that her approach could have been ambivalent; for while she lamented the low intellectual regard accorded to women when she had been young, and was no doubt aware of the still prevailing opinion which argued that women lacked men's capacity for *sustained* intellectual rigour, she nonetheless had a clear aversion to female dilettante intellectuals.

When the fledgling British Association for the Advancement of Science held its meeting in Oxford in 1832, for instance, it was generally felt that while ladies could attend the social side of the meeting with their husbands or fathers, they should be discouraged from attending the actual reading and discussion of the scientific research papers—otherwise their presence would turn the whole affair into an 'Albermarle (Royal Institution) soirée-dilettante meeting instead of a serious philosophical union of working men.' Therefore, would Mary—whose *On the Mechanism of the Heavens* had propelled her into international scientific celebrity —choose to attend the Association's meeting? In spite of encouragement from Dr. Somerville, Buckland, and others, 'In the end, Mrs. Somerville decided not to attend the meeting, for fear that her presence should encourage less capable representatives of her sex to be present.'[13] One senses that Mary was acutely aware of her unique and peculiar status in the scientific world, and feared that by failing to observe the accepted proprieties, she would ultimately do more harm than good.

On the other hand, throughout her life Mary was acutely aware of the discrimination against women in education, participation in public life, and elsewhere, and was ceaseless in her advocacy of women's rights. Yet what ultimately mattered to her was the promotion of intellectual excellence, and not just gender politics. Her attitude was that clever and highly motivated women should have opportunities to excel, and make their contribution in whichever branch of human endeavour their talents lay. In many respects, though, this was a fairly typical attitude of the age and level of

---

[12] John F.W. Herschel to Mary Somerville, 18 March 1844, in *Personal Recollections* [n.2], 265–268. Also, Mary Somerville to John Herschel, Rome, 12 November 1844, Royal Society MS, Herschel Papers HS16 [347], for further references to the Jesuit Observatory and its 'inaccessibility to women'.

[13] Mrs. (Elizabeth Oke) Gordon, *The Life and Correspondence of William Buckland, D.D., F.R.S.* (London 1894), 122–123. Also Elizabeth Patterson, *Mary Somerville and the Cultivation of Science, 1815–1840* (Martinus Nijhoff, Kluwer Group, Boston, The Hague, Lancaster, 1983), 192.

society in which she lived, whether the talents be possessed by men or by women. Intellectual culture in 1830 was, after all, fundamentally elitist, and believing as it did in progress, looked for new Newtons, Miltons and Beethovens to lead the way and keep raising the standards. Within the accepted standards of late Georgian and early Victorian Britain, Mary was very much aware of the fine line which she walked between that female radicalism or 'blue-stockingism' which she so disliked, and the conventional obedience of the model wife. Indeed, Dominique Arago, when visiting Dr. Thomas Young and his intellectual wife Eliza with Joseph Gay-Lussac in 1816, had highlighted this very point when he remarked that 'the fear of being designated *bas bleus* makes English ladies reserved in the presence of strangers'.[14]

Femininity was just as central to Mary's own personal identity as was creative mathematics, and she was always scrupulous to remain 'thoroughly and gracefully feminine' and avoid the general presumption that *thinking* women had ceased to be *real* women. Indeed, her friend Jane Marcet, writing from Geneva on 6 April 1834, described the essence of Mary's achievement, and what she had shown women to be capable of, when she said that with 'talents and acquirements of masculine magnitude, you unite the most sensitive and retiring modesty of the female sex'.[15] Likewise, her geologist friend Adam Sedgwick particularly commented on her femininity.

It might be argued that Mary's instinctive grasp of the rules of the social game was vital in enabling her to win such acclaim in what was still overwhelmingly a man's world. And as that society was primarily Grand Amateur in its basis, where good relations between individuals ultimately counted for more than relations with institutions, she was able to seek out her own opportunities and to flourish.

Over the course of Mary's long life, however, a fundamental sea-change took place regarding the wider role of women in Britain, continental Europe, and America. By the time that she entered her late seventies and eighties, she became acutely aware that she had played an influential part in bringing about that change, and she became keen to lend her support to publicly debated causes that would only have provoked laughter had they been raised when she was thirty. For example, she supported changes in the law which gave married women control of their own property in defiance of asset-stripping husbands, commenting: 'The British Laws are very adverse to women.'[16] And when the radical MP John Stuart Mill attempted

---

[14] In spite of Eliza Young's social caution not to appear a blue-stocking, when her husband, Arago and Gay-Lussac were discussing optical diffractions and the wave theory of light, she quietly left the room, only to return a minute later 'with an enormous quarto under her arm' (one of a series containing her husband's published researches). 'She placed it on the table, opened the book, without saying a word, at p. 387, and showed with her finger a figure where the curvilinear course of the diffracted bands which were the subject of the discussion, is found to be established theoretically': George Peacock, *Life of Thomas Young, M.D., F.R.S. and C.* (London, 1855), 388–389, taken from Arago's subsequent *Éloge* to Young.

[15] Jane Marcet to Mary Somerville (Geneva), 6 April 1834, in Mary Somerville, *Personal Recollections* [n.2], 209–210.

[16] Mary Somerville, *Personal Recollections* [n.2], 344.

to introduce amendments to Disraeli's Reform Bill of 1866 that would have given the vote to women, the 86-year-old Italian resident Mary Somerville gave him her full backing. Regarding Britain's laws, she said 'we are deeply indebted to Mr. Stuart Mill for daring to show their iniquity and injustices'.[17] Post-Civil War America, moreover, she considered equally unjust, for while as a lifelong abolitionist she rejoiced in the newly won freedom of the black slaves in the southern states, she still thought it absurd that Uncle Sam had immediately given the vote to its uneducated male liberated slaves, while 'refusing it to the most highly educated women of the Republic'.[18] And at the age of 89, she was quick to sign Mill's Parliamentary Petition. She was also quick to back up her convictions with a donation, for preserved in her papers in the Bodleian Library are financial documents itemising her donation of one guinea (£1.05) and a 10-shilling (50 p) donation from her daughters. And while this total of 31 shillings might appear rather small, one should not forget that in 1868 such a sum would have been equivalent to the butler's and cook's combined weekly wage for an English family (Fig. 5.3).[19]

One senses, however, that Mary's driving concern—in old age, just as when she had been a girl—lay in broadening access to education for women; for not until capable women were receiving the education which their talents merited would real progress be made in giving them a more responsible role in society. As she recorded: 'Age has not abated my zeal for the emancipation of my sex from the unreasonable prejudice too prevalent in Great Britain against a literary and scientific education for women.'[20] And one can see that as she approached 90 she was glad that her efforts, and the efforts of other advocates of women's education, were beginning to bear fruit. First there had been the Royal Holloway College for women; then, as she noted, Madame Emma Chenu had been awarded an M.A. degree in Paris; while a Russian lady had received a university degree. Mary had also signed an unsuccessful petition to the Senate of the University of London, 'praying that degrees might be granted to women'; she rejoiced at the founding of Girton College, Cambridge, in 1869;[21] and would have felt deeply proud to have had her name bestowed upon an Oxford women's college in 1879. She never ceased to deplore the enforced idleness in which most women of her class seemed willing to live, for amongst other things it occasioned such a ridiculous wastage of

---

[17] Mary Somerville had the greatest regard for John Stuart Mill, while Mill clearly regarded Mary as one of the most influential women of the age and a vital signatory to his Parliamentary Petition: J.S. Mill to Mary Somerville, 12 July 1869, *Personal Recollections* [n.2], 345. Martha Somerville added a note amplifying her mother's admiration for the 'noble character and transcendent intellect of Mr. J.S. Mill', in the passage included in *Queen of Science* [n.5], 277.

[18] Mary Somerville, *Personal Recollections* [n.2], 344.

[19] Receipt slip for one guinea and ten-shilling donations to Mary Somerville and her daughters from 'The London National Society for Women's Suffrage', 12 November 1868: Bodleian Library, Somerville Papers, Dep. c. 374 Folder MSBUS-12, S.C. Box 24.

[20] Mary Somerville, *Personal Recollections* [n.2], 345–346.

[21] Mary Somerville, *Personal Recollections* [n.2], 346–347.

Fig. 5.3 Mary Somerville. An undated engraving, *c.* 1845, bearing the legend 'From an original painting by Chappel in the possession of Johnson Wilson and Co., New York'. Might the original picture still survive in an American collection? (Author's collection)

talent that could otherwise have been employed in making the world a better place. In fact, this attitude was matched perfectly by that of her friend Lady Margaret Brodie Stewart Herschel (the wife of Sir John Herschel) who in 1869 lamented that 'In general ladies in India fall ill, for want of something better to do (and indeed it would be a good plan to give *them* appointments in India (along with their husbands) as part of the 'Women's Rights' Question).'[22]

Yet as Mary's daughter Martha Somerville recorded, when editing her mother's *Personal Recollections* in 1873, 'a commonly well-informed woman of the present

---

[22] Margaret B. Herschel to Mary Somerville, 14 April 1869: Bodleian Library Somerville MS. Dep. c. 370, Folder MSH-3. 34: file 42.

day would have been looked upon as a prodigy of learning in her youth'.[23] Great progress had therefore clearly been made during Mary's lifetime, and as women continued to enter into secondary and higher education, became school teachers, journalists, medical doctors, and writers, and forged the Suffrage Movement by 1900, their perceived role in society had become very different from what it had been a century earlier.

But virtually all of what we know about women in intellectual life in the nineteenth century relates to that class of persons who would have been considered 'ladies', and who had time, leisure and money readily available. I know of only a few cases of intellectually inclined women from the poorer classes of this period, and only one of these was known to Mary Somerville herself: Betty Veitch, the wife of the Scottish 'peasant' astronomer James Veitch.

All that we know of Betty Veitch, sadly, comes from a single incident which took place when Mary visited the Veitch farm at Inchbonny, Jedburgh. Betty Veitch 'seemed to be a person of intelligence,' Mary recorded, 'for I remember seeing her come from the washing tub to point out the planet Venus while it was still day-light'.[24] This was no mean achievement, and virtually impossible for anyone not intimately familiar with the movements of the heavens.

Another working-class woman who seems to have had intellectual interests was also a Scot: the wife of John Robertson, a porter at Coupar Angus station, near Perth. In 1884 John Robertson was interviewed by the biographical writer Samuel Smiles, as a result of his local prominence as an amateur astronomer in 'humble life'. Unfortunately, not even Mrs. Robertson's Christian name was recorded by Smiles, who met and was entertained by her in their cottage at Causewayend Street, Coupar Angus, although we are told that she was 'evidently clever'.[25]

A rather less shadowy wife of an artisan astronomer was Anne Langdon, wife of Roger, who also earned his living as a porter, and after 1867, as a village Station Master, on the Great Western Railway at Silverton, near Exeter, Devon. Anne Langdon not only shared her husband's scientific interests, but also his theological and literary interests, and both of them were entirely self-educated. At Silverton, moreover, Anne began to run an unofficial village school in the Station Master's house, and played an active part in the spiritual and mental life of that rural Devonshire community. From the Life of Roger Langdon—which their daughter Ellen edited from family documents as a memorial to her parents in 1909—Anne and Roger Langdon seem to have enjoyed a remarkably equal relationship, and

---

[23] Martha Somerville's note: Mary Somerville, Personal Recollections [n.2], 346. Several paragraphs regarding women's rights, education and professional prospects that were omitted from the published Personal Recollections were included by Dorothy McMillan in Queen of Science [n.5], 277–280.

[24] Mary Somerville, Personal Recollections [n.2], 100. Mary does not mention Mrs. Veitch's Christian name, although her maiden name had been Betty Robson: cited without primary source by J.N. McKie, 'James Veitch 1771–1838', Journal of the British Astronomical Association, 87, 1 (1976), 44–50, from G. Watson, The Border Magazine, V (1), January 1900.

[25] Samuel Smiles, Men of Invention and Industry (London [3]), 328.

Roger himself openly admitted that without her support his own scientific and theological studies would have been impossible.[26]

How many similar women there might have been—who kept a family and bought second-hand books on a pound or thirty shillings a week—we do not know. Unlike their more affluent sisters in the Grand Amateur classes, these women, like their husbands, did not send letters that came to be preserved in major archives; neither did they write books, nor receive mention in the newspapers. What we know of figures like Betty Veitch, Mrs. Robertson and Anne Langdon is preserved only in isolated scraps. Unfortunately, there appears to be no evidence that any such women wrote letters to Mary Somerville.

We have seen that Mary was firmly associated throughout most of her life with progressive and humanitarian causes such as the abolition of slavery, women's rights, progressive science and political reform. Likewise, most of her friends belonged to the Whig or Liberal end of the spectrum. However, her allegiance to the geologists, sympathy with the outrageous *Vestiges* in 1844, guarded support for Darwin's *On the Origin of Species* after 1859, and being preached against in York Minster, all pose questions about Mary's religious beliefs.

Early Victorian religion had become a minefield, replete with conflicts and contradictions.[27] The overall religious tenor of England, Scotland, Wales and the politically influential parts of Ireland was low-church Protestant. Bible-based in its Christian expression both within Anglicanism and the Protestant dissenting churches, the spiritual fabric of British society, and especially Oxford University, was torn asunder in the 1840s in the wake of John Henry Newman's Oxford Movement. Newman and his friends John Keble, Edward Bouverie Pusey and others advocated an opening up of the Church of England to the historical Christianity of Roman Catholicism. For a country which, over 300 years, had become accustomed to seeing the Church of Rome as Antichrist, the Oxford Movement struck one of the most sensitive nerves in the English body politic. Keble and Pusey remained within the Anglican Church, but in 1845 Newman resigned his Vicarage of Oxford's University Church to go to Rome, and became a Catholic priest and later a Cardinal. Over the years, Newman's powerful spiritual writings encouraged many clever young men and women to follow him into the hated Church of Rome.

In addition to internal doctrinal fights between Anglicans, Catholics, Methodists, Presbyterians, Unitarians and others, there were yet further intellectual assaults upon traditional Christian society emanating from continental Europe. David Friedrich Strauss's *Das Leben Jesu* (1835) horrified people as it attempted to advance a critical biography of Christ, as opposed to the more familiar reverential treatments of His life. Other continental exponents of 'Higher Criticism' were beginning to analyse Old Testament stories in the light of parallel studies of the

---

[26] Roger Langdon, *The Life of Roger Langdon told by himself, with additions by his Daughter Ellen* (London [4]), 65–6.

[27] The most comprehensive, and most readable, history of religion in early nineteenth-century Britain is Owen Chadwick, *The Victorian Church, Part I, 1829–1859* (S.C.M. Press, 1966).

other cultures of the Near East, especially after Young and then Champollion had deciphered the Egyptian hieroglyphic script during the 1820s. Did Bible stories have cognates in Greek mythology, or even in the old European folk tales being collected by the brothers Jacob and Wilhelm Grimm? And from the 1830s onwards, the secular Positivist philosophy of Auguste Comte was advocating the scientific method as the only source of true or 'positive' knowledge.

To think of the fossil geologists, *Vestiges,* or Darwin as somehow upsetting the hitherto placid spiritual landscape of Britain is to miss the big picture of religious turmoil that was already a foot. And while the Christian faith would soon learn how to handle these threats, move away from a narrow Biblical literalism, absorb important aspects of Catholic spirituality, and regard evolution as a more complex form of God's providence, not all of these developments were obvious in Mary's lifetime, and many thinking Victorians felt shocked and confused by the apparent undermining of traditional forms of Christian belief.[28]

While Mary Somerville was acutely switched on to the intellectual currents of her day, there is not really any evidence that they occasioned any fundamental turmoil in her own personal life. Perhaps more than anything else, this derives from that natural cool-headedness and lack of emotionalism in her make-up that has already been noted. There is no sign in her writings that she, unlike Darwin and so many of her contemporaries, ever agonised about hell fire or the expiation of sin.

On the other hand, Mary was in no way indifferent to religion, for as her daughter Martha later recorded, she was 'profoundly and sincerely religious'.[29] But she had a dislike of creeds and dogmas which she felt were as much coloured by human wishes to control as they were by God's grace, for 'hers was not a religion of mere forms and doctrines, but a solemn deep-rooted faith which influenced every thought, and regulated every action of her life'.[30] It was also an approach to religion that was innocent of bigotry, and in which she felt glad, in spite of her Protestantism, to be presented to two Popes (as mentioned in Chap. 2). Furthermore, it was also quite clear that she reckoned a Pope purely in terms of his personal qualities, and not as Christ's Vicar on Earth; and her northern Protestantism also shone through when describing particular acts of Catholic devotion undertaken by the local peasants, such as when the Aurora Borealis—visible in Naples in 1870—led to the locals praying to the Madonna for deliverance.[31]

---

[28] Chadwick, *The Victorian Church, Part I* [n.27], 527–572. Owen Chadwick, *The Victorian Church, Part II, 1860–1901* (S.C.M. Press, London 1972) deals at length with these new intellectual problems in Chapters I, II and III. Also R.J. Berry, 'Evolution', in *The Oxford Companion to Christian Thought,* ed. Adrian Hastings, Alistair Mason and Hugh Piper, with Ingrid Lawrie and Cecily Bennett (Oxford University Press 2000), 224–226.

[29] Mary Somerville, *Personal Recollections* [n.2], 374, 376. Religious as she undoubtedly was, Mary Somerville had nothing but contempt for superstition, and in an earlier draft omitted from the published *Personal Recollections* she indicated her scorn for the Victorian craze for spiritualism and 'table rapping': Dorothy McMillan, *Queen of Science* [n.5], 277.

[30] Mary Somerville, *Personal Recollections* [n.2], 374.

[31] Mary Somerville, *Personal Recollections* [n.2], 353.

It is unclear whether Mary thought in terms of a close personal relationship with
God—which was often a source of inner turmoil for many of her more traditionally
Christian contemporaries—or, alternatively, of the joyous beholding of a glorious
Providence in action. What cannot be denied, however, is that a religious delight
runs through her books. *On the Connexion of the Physical Sciences,* for instance,
draws all of its complex strands of argument together into an overtly religious
conclusion. Not only does the unity of science point to a Grand Design, but so too
does the intellectual constitution of the scientific method itself: 'This mighty
instrument of human power itself originates in the primitive constitution of the
human mind, and rests upon a few fundamental axioms, which have existed eternally in Him who implanted them in the breast of man when he created him after
His own image'.[32] Indeed, one might argue here for something far more specific
than the General Providences referred to by eighteenth-century deist writers, for
Mary is stating quite unequivocally that the logical self-awareness of human beings
from which science springs is itself the direct result of a gift from, and a special
relationship with, the Creator of all things.

Her religious faith, therefore, seems to have been an non-dogmatic Christianity
that was grounded in mankind's unique relationship to the same 'Almighty' whose
hand had shaped the planetary orbits, the terrestrial continents, and the human mind.
It is true that she says nothing about how she interpreted the deaths of several of her
children, two husbands, and many friends, and a world of suffering in the light of this
broad beneficence; though as she recorded at the age of 89, she contemplated
entering 'that new state of existence' calmly trusting in God. Judgment, purgatory,
hell and punishment do not seem to have clouded her spiritual horizons. Instead, she
seems to have seen no reason why a loving God should not finally draw all of His
creatures back to Himself, especially if they had lived virtuous lives and done good
to their fellows.

What is more, this seems to have been an approach to life, death and immortality
which one also encounters in the writings of several of her friends. The 83-year-old
Adam Sedgwick (who was 5 years Mary's junior)—finally emerging from his
bachelor rooms in Trinity, Cambridge, in mid-April 1869, after 4 months of sickly
confinement due to vertigo, bronchitis, 'suppressed gout' and an eye inflammation
—could lament without too much anguish the deaths of old friends such as William
Whewell—the 'last of the old stock'—before conferring his cheerful blessings on
the 89 year old Mary. 'May God preserve and bless you!', wrote Sedgwick, 'and
when so ever it may be His will to call you away to Himself, may your mind be
without a cloud, and your heart full of joyful Christian hope'.[33]

But some of her most revealing correspondence, as far as religion is concerned,
is that exchanged with the obviously fading Sir John Herschel, and his family,

---

[32] Mary Somerville, *On the Connexion of the Physical Sciences,* 3rd edn. (London, 1836), 412.
These words concluded all of the editions of *Connexion.*

[33] Adam Sedgwick to Mary Somerville, 21 April, 1869, in J. W. Clark and T.M. Hughes, *The Life
and Letters of the Reverend Adam Sedgwick, LL.D., D.C.L., F.R.S.,* II (Cambridge University
Press, 1890), 446. Sedgwick's letter is reprinted in *Personal Recollections* [n.2], 365–6.

between the late 1860s and Herschel's death in 1871. As usual, of course, their letters contain an abundance of scientific exchange—about the 1868 eruption of Vesuvius,[34] the absolute temperature of space as extrapolated from high-altitude balloon ascents,[35] and meteorological and barometric observations[36]—and also Herschel's translation of Dante.[37] As the person who had always cast a critical eye over manuscript drafts of her books, Herschel was requested to read the first versions of her future *Personal Recollections*. In this capacity, moreover, Herschel was quick to correct Mary's treatment of the discovery of Neptune in 1846, where in her manuscript draft she had depicted John Couch Adams as a brilliant young mathematician (which he undoubtedly was) who had lost the priority for discovering Neptune to the French because of the tardiness of the Greenwich and Cambridge Observatories. But as Herschel—who lived through all of the repercussions of the discovery in 1846 when Mary had been residing in Italy—reminded her, the apportionment of praise and blame in the affair was not straightforward. Indeed, said Herschel: 'It is, of all points in the history of Astronomical discovery that which most needs wary walking to do strict Justice—and you cut through the Gordian knot with a slash.' Likewise, in the same letter, Herschel advised her to avoid discussing the complicated business of priority between Cooke and Wheatstone in the invention of the electric telegraph.[38] In both cases, Mary took Herschel's advice, and radically modified her original draft to that which would eventually go to the printer.

In Mary Somerville's last correspondence with Sir John Herschel—via Lady Margaret, his wife—there are increasing clues to his failing health. By 1869, said Lady Herschel: 'His walk and gait are decidedly feeble and bent but his Eye lights up wonderfully at the mention of any Kindred theme or fond memory.' Yet the end of 'our Philosopher gliding down the Evening of his life with the glory of a Setting Sun' was awaited cheerfully and with no apparent terror. And when Herschel did pass away, on 11 May 1871, his wife characterised him as having *'gone Home, home to the Father who lent him to us for a while, and who will now perfect him to*

---

[34] Mary Somerville to Sir John F.W. Herschel, Naples, 12 November 1868, Royal Society, Herschel Papers HS16 [375].

[35] Sir John F.W. Herschel to Mary Somerville, 20 January 1858, Royal Society, Herschel Papers HS16 [360].

[36] Maty Somerville to John F.W. Herschel, 23 October 1866 ('Spezia'), Royal Society, Herschel Papers HS16 [374].

[37] Mary Somerville to John F.W. Herschel (Naples), 26 June 1868, Royal Society, Herschel Papers HS1.6 [376].

[38] John F.W. Herschel to Mary Somerville, 14 March 1869, Royal Society, Herschel Papers IIS16 [378]. For the controversial passage about Adams and the 1846 discovery of Neptune which, upon Herschel's advice, was struck out of Mary Somerville's published *Personal Recollections*, see the manuscript Bodleian Library, Somerville MS. Dep. c. 355 Folder MSAU-2, p. 65, in the original hand, but refoliated in pencil, 221. A milder version of the passage was composed, and appears in the same Bodleian Somerville MS., 220 verso. The version which eventually appeared in *Personal Recollections* [n.2], 290, was shorter, and less judgmental of those English astronomers who were supposedly responsible for not acting upon Adams' calculation.

do still 'Greater things than these'.[39] Here one finds an attitude to life, death and religion which was very similar to that of Mary herself: a good and useful life which, after its termination, expected neither hell nor terror, but rather a perfecting of life's infirmities, followed by yet more glorious wonders to behold and, perhaps, the chance to speak with the Almighty Architect who had first set the planets spinning in their courses.

Optimistic as she clearly was of a future reunion in Heaven with her husband, with Herschel, and with all of their old friends, it is plain that the passing of Herschel in 1871 brought home to Mary a sense of relative isolation in this world, and of having outlived so many of her family and friends. She was 'deeply grieved and shaken by the death of Sir John Herschel, who though ten (actually twelve) years younger than I am, has gone before me. In him I have lost a dear and affectionate friend.'[40] Indeed, this passage is one of the relatively few glimpses she affords us of how deeply the deaths of many family members and friends must have affected her.

Towards the end of her autobiographical draft, as she contemplated her own death, she admitted that after a long, sometimes trying, but essentially happy life, 'I think of death with perfect composure and perfect confidence in the Mercy of God.'[41] Although she regretted leaving such a beautiful and fascinating world, and knew that she would never hear of the oceanographic findings of the HMS *Challenger* expedition (1872–1876), nor see the transits of Venus in 1874 and 1882, her greatest regret was that she had not lived to see slavery abolished in Africa. But she was an Admiral's daughter to the last, and with the sea in the blood of Fairfaxes, Greigs and Somervilles for five generations, she characterised her own impending end in nautical terms—as being similar to that of a warship awaiting its departure signal: 'The Blue Peter has long been flying at my foremast, and now that I am in my 92nd year I must soon expect the signal for sailing. It is a solemn voyage, but it does not disturb my tranquillity.'[42]

At the age of 92, Mary Somerville was still in full possession of her intellectual faculties (though she admitted to deafness and a tendency to forget personal names) and capable of understanding mathematics. On 28 November 1872 she had been correcting proofs; then, in the morning of 29 November she received her signal to sail, and slipped quietly out of harbour while in her sleep. She was buried in the English Cemetery in Naples, with her husband.[43]

---

[39] Margaret B. Herschel to Mary Somerville, 28 May 1871, Bodleian Library Somerville MS. Dep. 3. 370 HSH3, Folder 42.

[40] Mary Somerville, *Personal Recollections* [n.2], 361.

[41] Printed by Elizabeth Patterson, *Mary Somerville 1780–1872* (Oxford, 1979), 44. Dr. Patterson says that the passages comes from 'towards the end of her long autobiographical manuscript'; however, I was not able to find it in the Bodleian Library, Somerville Papers, Dep. c. 355, MSAU-2 or 3 (although this was probably an oversight on my part).

[42] Mary Somerville, *Personal Recollections* [n.2], pp. 373–374.

[43] Martha Somerville's note: Mary Somerville, *Personal Recollections* [n.2], p. 377.

References

1. Elizabeth, P. (1979). *Mary somerville 1780–1872* (p. 33). Oxford: Oxford University Press.
2. Moseley, M., & Genius, I. (1964). *A life of Charles Babbage* (p. 167). London: Hutchinson.
3. Smiles, S. (1884). *Men of invention and industry* (p. 328). London: John Murray.
4. Langdon, R. (1909). *The life of roger langdon told by himself, with additions by his daughter ellen* (pp. 65–66). London.

Chapter 6
Conclusion: A Career in Retrospect

Would Mary Somerville have achieved more, and gone on to make major scientific discoveries, had she been able to go to university, hold down a job, and become the Professorial Director of some research institute? Richard Anthony Proctor, her Royal Astronomical Society obituarist in 1872, lamented that the social proprieties of her day had made it impossible for Mary to really achieve her potential as a planetary dynamicist, while Dr. Mary Brück, in her excellent and perceptive biographical article, styled her a 'mathematician and astronomer of underused talents'.[1] It is difficult to deny, of course, that had Mary Somerville been born into a later generation, gone, let us say, to Girton or Somerville College, had access to computers, and ascended an accessible promotion ladder in the world of physics, the nature of her contribution would have been radically different. Most notably, it would probably have focused much more upon scientific discovery rather than upon scientific communication.

Yet Mary Somerville was a product of the late eighteenth century, and it is not realistic to consider her talent simply as abstract potential that might theoretically be planted into a selected time frame and its growth therein imagined. Had she lived in the twelfth century, for instance, would Mary have been a famous theologian Abbess? Or had she been born in the twentieth century, would she have become a Nobel Laureate in Physics? It is true that if she had been born a man in 1780, her career would almost certainly have been totally different—and even if this had precluded a career in the Royal Navy, and perhaps death in action during the Napoleonic wars, a career in academic science as we now think of it was still by no means probable. For as we saw in Chap. 1, those scientific chairs that existed in the English, Scottish and Irish universities were notoriously poorly paid, and presupposed that their incumbents would at least in part finance their own research, and even pay for capital equipment out of their own pockets. This was the world of the Grand Amateurs, in which status and distinction were invariably linked to personal

---

[1] See Ref. [1, 2]

© The Author(s) 2015
A. Chapman, *Mary Somerville and the World of Science*,
SpringerBriefs in History of Science and Technology,
DOI 10.1007/978-3-319-09399-4_6

independence. Yet it was in this Grand Amateur world that Mary found her scientific voice and established her reputation.

Had Mary Somerville been born a man in 1780, yet lacked that secure inheritance to launch upon a Grand Amateur career in science—which the straitened circumstances of the Fairfaxes would have made probable—then that individual would have had to earn a living. One of the ways in which this could have been done was to have been a scientific author—just like Proctor himself. For while Proctor lamented 'what, under happier auspices, she might have accomplished' (presumably by having the opportunity to hold a scientific post) in his obituary essay on Mary Somerville, one should not forget that his own astronomical career had been very similar to Mary's. It was not a chair or observatory directorship which provided the foundation for his extremely successful and lucrative astronomical career, but the authorship of several dozen best-selling scientific books. However, one also detects Proctor's personal regret in being obliged to earn his living thus, and in his *Wages and Wants of Science Workers* (1876)[2] he castigates the government for not fostering the employment and proper payment of scientists.

Yet Proctor was by no means unique in his no doubt inadvertent following of Mary Somerville down the path of scientific authorship when ample private funds were not available. A similar path was taken by the Irish historian and interpreter of contemporary astronomy, Agnes Clerke, who was not only acutely aware of Mary's influence on the scientific community of her day, but whose sister, Ellen Mary Clerke, wrote the entry on Mary in the *Dictionary of National Biography* in the mid-1880s. From the early 1880s down to her death in 1907, Agnes Clerke came to enjoy a formidable reputation as an interpreter of the astronomical developments of the nineteenth century—especially astrophysics—and enjoyed friendships with Sir David Gill, Sir William and Lady Huggins, and the Astronomer Royal, Sir William Christie. She learned practical, big-telescope astronomy at the Royal Observatory, Cape of Good Hope, and was able to decline a not especially well-paid post on the staff at Greenwich because of the greater means and freedom that her scientific writing earned for her. But as we have seen elsewhere in this book, private means, a profession or authorship was often the way in which British scientists of both sexes funded their work during the Georgian and Victorian ages, in contrast with the selective state patronage which existed in Paris, St Petersburg or Berlin.

With all of these factors duly considered, however, there was still one major circumstance in Mary Somerville's career which would have been very different had she been a man. Dr. Somerville would not have been obliged to visit libraries to

---

[2] Proctor, 'Mary Somerville', obituary [n. 1], 196, argues against the lack of official funding, especially for salary provision and adequate posts, for British scientists. Even so, he admitted that a successful scientific writer could make between £2,000 and £5,000 a year from writing, while a *new edition* of a successful work could earn for its author as much as the entire annual salary of a university professor: '(say from £500 to £1,000)', p. 8. While I am not aware of any precise figures for Mary Somerville's earnings from her books—especially from her consistently in print *On the Connexion of the Physical Sciences* and *Physical Geography*—one might take Proctor's figure from the 1870s as a rough guide to *potential* earnings for 20 or 30 years earlier.

transcribe sections of books on her behalf, nor would he have had to visit the Collegio Romano Observatory for her, act as her channel of correspondence with the world of scientific men, or fulfil the formal duty of chaperone when she visited Parisian and other scientific gatherings. Had she been a man, Mary would also have been able to be a full and participating member of the Royal Astronomical Society, the Royal Society, the Royal Irish Academy, or any other learned society. Nor would there have been any equivocation about attending paper-reading sessions at the British Association.

As we have said already, it is the historian's task to take the past on its own terms, and when one does this, Mary Somerville's career assumes a remarkable significance. Not only could her particular career have flourished only in a Grand Amateur scientific environment in 1830, but she used her prominence and intellectual authority to advance the cause of women in society at large. Not only was an Oxford college named in her honour in 1879, but after the publication of *Mechanism* in 1831 she was actively showing the world what women were capable of doing in science. Neither should her influence upon the nineteenth century be underestimated, for unlike the deliberately retiring Caroline Herschel, Mary Somerville was a *public* figure. And while at first she may have been venerated as a curious prodigy rather than as one of an established type, it cannot be denied that by 1900 the path which she illumined had certainly been followed by an increasing number of other women of varying talents. It is also true that while scientific and intellectual women were still being laughed at and mocked in some quarters—even in the early twentieth century—this in itself was an indication that a 'type' rather than a few isolated individuals had most conspicuously come into being by that time.

As the Grand Amateur element gradually slipped from the forefront of British science by 1900, however, new types of scientific organisation were being founded to cater for those individuals who wished to pursue astronomy and other sciences—not in the expectation of making great discoveries, but as a serious hobby. Amateur astronomical societies, their memberships consisting of local solicitors, family doctors, schoolteachers and businessmen, were appearing in many of the great cities of Britain by the 1890s, and in 1890 itself the British Astronomical Association was founded in London, to coordinate and meet the needs of amateur astronomers across Great Britain and her dominions. Furthermore, all of these new amateur astronomical societies became immediately open to women. The founder members of the British Astronomical Association included Miss Mary Orr (later Mrs Mary Evershed), and Miss Elizabeth Brown of Cirencester—who, along with her unmarried sister Jemima, pursued her scientific and intellectual activities on the strength of the inherited profits of a family wine-importing business—and on its very first governing council in 1890 the Association had three women, two of whom were university graduates.[3]

---

[3] In Memoriam see Ref. [3, 4]

By the early twentieth century there were more than a hundred women members on the rolls of the British Astronomical Association and the provincial astronomical societies across Britain; and there were yet more women members in those similar amateur scientific societies devoted to natural history, geology, meteorology and archaeology. But not all of them were *amateurs* in the proper sense. Growing numbers of these women were university graduates and had paid jobs in grammar schools, universities and research institutes, but saw the amateur societies as an important venue for female intellectual association. Then, in 1916, after 50 years of discussion, the Royal Astronomical Society finally opened its portals to women on the same terms as those enjoyed by men, some 81 years after granting special *Honorary* Membership to Caroline Herschel and Mary Somerville (and later to Lady Margaret Huggins).

By the time that Florence Taylor delivered her lecture on the career and achievements of Mary Somerville to the approving and applauding membership of the Leeds Astronomical Society in 1897, all were aware that a fundamental change had taken place in European and American society as far as the status of women and their education was concerned. And in bringing about that change, Mary Somerville herself played a very major part.

Reference

1. Richard, A. P., (1873). Mary Somerville, Obituary Notice. *Monthly Notices of the RoyalAstronomical Society*, *33*, 190–197.
2. Mary, B., (1996). Mary Somerville, mathematician andastronomer of underused talents. *Journal of the British Astronomical Association*, *106*, 4,201–206.
3. Elizabeth, B., (1899). F.R. Met. Soc.', obituary. *Journal of the British AstronomicalAssociation*, *9*, 5, 214–15.
4. Chapman, A., (1998)The Victorian Amateur Astronomer. Independent Astronomical Research in Britain, 1820–1920 (pp. 280–293). Chichester, New York: Praxis-Wiley.

Index

© The Author(s) 2015
A. Chapman, *Mary Somerville and the World of Science*,
SpringerBriefs in History of Science and Technology,
DOI 10.1007/978-3-319-09399-4